OF EMPIRE AND THE CITY

NEW STUDIES IN EUROPEAN CINEMA

VOL. 15

EDITED BY
WENDY EVERETT & AXEL GOODBODY

Peter Lang
Oxford • Bern • Berlin • Bruxelles • Frankfurt am Main • New York • Wien

Maurizio Cinquegrani

OF EMPIRE AND THE CITY

Remapping Early British Cinema

Peter Lang
Oxford • Bern • Berlin • Bruxelles • Frankfurt am Main • New York • Wien

Bibliographic information published by Die Deutsche Nationalbibliothek
Die Deutsche Nationalbibliothek lists this publication in the Deutsche Nationalbibliografie; detailed bibliographic data is available on the Internet at http://dnb.d-nb.de.

A catalogue record for this book is available from the British Library.

Library of Congress Control Number: 2013957634

Cover image: Charles Ginner, *Piccadilly Circus*, 1912, oil paint on canvas

ISSN 1661-0261
ISBN 978-3-0343-0823-6

Hochfeldstrasse 32, CH-3012 Bern, Switzerland
info@peterlang.com, www.peterlang.com, www.peterlang.net

This publication has been peer reviewed.

Contents

Acknowledgements vii

INTRODUCTION
The Absent-minded Cameraman? 1

PART I London 41

CHAPTER 1
Film at the Heart of the Empire 43

CHAPTER 2
Connecting Places: Film, Transport and Display 85

PART II The British Isles 113

CHAPTER 3
Cinematic Empire: Seaside, Industry and Trade 115

CHAPTER 4
Regional Urban Spaces: Scotland in Film 147

CHAPTER 5
Hybrid Space: The Irish City 165

PART III The Empire 185

CHAPTER 6
Travel Cinematography in India 187

CHAPTER 7
The Perilous Exoticism of the East 217

CHAPTER 8
From the Boer War to the Great War 227

Bibliography 249

Filmography 275

Index 297

Acknowledgements

I would like to thank the Arts and Humanities Research Council (AHRC) for supporting my research with a Doctoral Award. My supervisors, David Green and Mark Shiel, have given me enormous help during my research. Thanks.

I should also thank Palgrave Macmillan, Tate Britain, the University of Edinburgh Press, and Taylor & Francis for the permission to include here sections of the articles and book chapters I have authored for these publishers. Portions of Chapter 1 have been published previously as "Empire and the City: Early Films of London" (*The Camden Town Group in Context*, eds Helena Bonett, Ysanne Holt, Jennifer Mundy, London: Tate Britain, 2012) and "The Nexus of the Empire: Early Actuality Films of London at the Turn of the Twentieth Century" (*Journal of British Cinema and Television*, vol. 6, no. 1, 2009, pp. 207–219). Sections of Chapter 2 have been published previously as "The Cinematic Production of Iconic Spaces in Early Films of London" (*The City and the Moving Image: Urban Projections*, eds Richard Koeck and Les Roberts, Houndmills, Basingstoke: Palgrave Macmillan, 2010, pp. 169–182). Portions of Chapter 6 have been published previously as "Travel Cinematography and the Indian City: Imperial Spectacle at the End of the Long Nineteenth Century" (*Nineteenth-Century Contexts: an Interdisciplinary Journal*, vol. 3, no. 1, 2010, pp. 65–78). Parts of Chapter 8 have been published previously as "'A fit of absence of mind'?: Empire and Urban Life in Early Non-Fiction Films" (*Early Popular Visual Culture*, vol. 9, no. 4, 2011, pp. 325–336).

Ideas, support, help and inspiration have been given generously by friends and colleagues, among them Jon Burrows, Bryony Dixon, Sarah Cooper, Patrick Keiller, Richard Koeck, John MacKenzie, Lynda Nead, Sally Parrott, François Penz, Laurel Plapp, Vanessa Toulmin and Lawrence Webb. I also wish to thank my students, whose thoughtful discussion of the ideas in this book has enabled its content to grow.

Above all I wish to thank my family.

INTRODUCTION

The Absent-minded Cameraman?

> We seem, as it were, to have conquered and peopled half the world in a fit of absence of mind.
>
> — SIR JOHN SEELEY, *The Expansion of England* (1893)

> The world of today is a bare, hungry, dilapidated place compared with the world that existed before 1914, and still more so if compared with the imaginary future to which the people of that period looked forward.
>
> — EMMANUEL GOLDSTEIN, *The Theory and Practice of Oligarchical Collectivism*

A middle-class man born in London on 20 June 1837, Queen Victoria's coronation day, could have been lucky enough to visit the Zoo in Regent's Park as a young boy after its opening to the public in 1847. He could have also visited the Crystal Palace in Hyde Park or James Wyld's Globe in Leicester Square in his early teens. As he turned twenty he would have read about the Sepoy Mutiny in India, perhaps in one of those illustrated magazines which often devoted their pages to events taking place in the colonies. Later he could have read serialized novels by Charles Dickens in *All Year Around* or Arthur Conan Doyle's short stories on *The Strand* while sitting on a train boarded in one of the fourteen major railway stations which opened in London during his lifetime. The train could have taken him on a short holiday to a seaside resort like Brighton or Blackpool, or visiting relatives in Manchester, Glasgow or Liverpool. Unless he had joined the army or the civil service and travelled to distant destinations, his knowledge of the colonies would have come from shows, exhibitions, books, postcards, and magazines. At the age of sixty he would have hardly been able to dissimulate his wonder in front of the first exhibitions of moving pictures in his local music hall. Perhaps too tired to travel to central London or unwilling to face the great crowd, he would have watched the Diamond Jubilee procession on film. He would later get used to the new technology, as he would realize that the themes and representational strategies of films were already familiar: exhibitions, means of transport, parades, crowded street and images of distant and exotic places had long been part of his experience of urban life in Victorian Britain.

This age witnessed the arrival of mass communication and mass visual culture, as new technologies of printing, reporting and reproduction fuelled an unprecedented expansion in the circulation of material such as newspapers, magazines and photographs. Towards the end of the Victorian era, technological advances aimed at capturing images and motion on a light-sensitive medium led to the invention of film. Early cinema, in Britain and everywhere else, was a quintessentially urban phenomenon and city life was a favourite subject for audiences and filmmakers.

Images of cities in visual culture, previously the subject of paintings, prints, and novels, became increasingly accessible in the nineteenth century because of the possibilities given by photographic reproduction. Cinematography was part of a wider series of challenges to that uniform notion of perspectival space which had governed visual culture since the Renaissance and which, as Stephen Kern has argued, was being questioned by artists, novelists, scientists and sociologists for the first time. According to Kern, cinema offered new spatial possibilities and displayed distant and exotic places to which the audience rarely had access, within the context of a series of dramatic changes in technology and culture leading to new ways of thinking about and experiencing time and space.[1] In this book, I investigate the cinematic experience of urban space at the turn of the twentieth century and the historical and political conditions which defined it. These conditions included the new technologies which began to penetrate everyday life at the end of the nineteenth century when, according to Henri Lefebvre, the contours of modernity emerged vividly. Progress and

1 Stephen Kern, *The Culture of Time and Space: 1880–1918* (Cambridge, MA and London: Harvard University Press, 2003; first published in 1983), 1–6, 142–143.

bourgeois well-being, Lefebvre suggests, was short-lived, as the escalation of the technological progress was soon applied to armaments, the threat of war and eventually war itself.[2] Nevertheless, at the turn of the twentieth century, a certain democratization of visual culture, technological progress, and improved living and working conditions defined the predictive vision of a rich, leisured, orderly and efficient future society which, according to George Orwell and through the words of Emmanuel Goldstein, was expected by the literate masses. Goldstein, whose quote opens this book, is a fictional character in George Orwell's *Nineteen Eighty-Four*. In the novel, he is either a former top member of the ruling Party who had broken away early in the movement and started an organization dedicated to the fall of the Party or, perhaps, a scapegoat fabricated by the Party itself. *The Theory and Practice of Oligarchical Collectivism* is a fictitious book allegedly written by Goldstein and secretly read by Winston Smith, the main character in Orwell's novel.[3] This "imaginary future" expected by people in the West had its roots in the technological progress and middle-class well-being which characterized the late nineteenth and early twentieth century.

This book looks back at that period and explores cinematic cities of the British Empire during the years between the invention of cinematography in 1895 and the outbreak of the First World War. This period coincides with the second half of the era defined by Eric Hobsbawm as the "Age of Empire".[4] Between 1875 and 1914, the capitalist core of the world economy, according to Hobsbawm, contributed to the creation of a world of empire where large numbers of people around the world were dominated by technologically and industrially advanced countries. During those years, the mass market and technological change contributed to the spread of the imperial message and, Hobsbawm argues, revolutionized the arts appealing to the working classes. The official birth of cinema thus coincided with

2 Henri Lefebvre, *Introduction to Modernity*, trans. by John Moore (London and New York: Verso, 1995; first published as *Introduction à la modernité*, Paris: Les Editions de Minuit, 1962), 178–179.

3 George Orwell, *Nineteen Eighty-Four* (London: Penguin) 191–226 (196).

4 Eric Hobsbawm, *The Age of Empire: 1875–1914* (London: Abacus, 2006; first published by Weidenfeld & Nicolson, 1987), 220, 337–338.

the height of British imperialism.[5] The outbreak of the First World War in 1914, on the other hand, coincided with radical changes in film production, the application of technological novelties to strategies of annihilation and the end of the long-enduring nineteenth-century balance of power.[6] The aim of this book is to explore the ways in which the city on film came to signify the historical achievements of a Victorian Britain. Most films from this period are now gone: they have disappeared, burnt, or destroyed. Many of those which have survived are scratched and ruined, but they still offer an invaluable record of city life at the turn of the twentieth century. This book is also based on primary and secondary records of long-lost films; this material contributes to offering a more precise definition of the themes, places and spaces emerging from early non-fiction cinematography.[7]

The intersection between recent studies of the cinematic city, scholarly work on early cinema, and the cultural history of the British Empire is the junction where I aim at tracing an interdisciplinary discourse on early films of imperial and colonial cities. This book brings together topics and studies not previously united, and promotes a new understanding of the early cinematic city. On one hand, I propose a re-evaluation of the often neglected practice of early non-fiction filmmaking and, on the other, I aim at assembling a narrative of imperialist influences on British cities and their cinematic representation. In this way, I want to argue for a shift in focus in the study of early cinema. Arising from the intertwining of cultures of modernity, as Vanessa Schwartz suggests, non-fiction films marked an

5 Hobsbawm, 338–339.

6 In the mid 1910s feature-length fiction films established themselves as the dominant form of cinema. On the birth of the Institutional Mode of Representation, see Noël Burch, *Life to Those Shadows* (London: BFI Publishing, 1990). On the end of the nineteenth-century balance of power see Hobsbawm's works: *The Age of Revolution: Europe 1789–1848* (London: Weidenfeld and Nicholson, 1962), *The Age of Capital: Europe 1848–1875* (London: Weidenfeld and Nicolson, 1975), and, in particular, *The Age of Empire: Europe 1875–1914* (London: Abacus, 2006. First edition published by Weidenfeld & Nicolson, 1987).

7 The cataloguing work of Denis Gifford has been particularly helpful in creating a comprehensive filmography. See Denis Gifford, *The British Film Catalogue: Non-Fiction Film, 1888–1994* (London: Fitzroy Dearborn, 2001, 3rd edn).

unprecedented crossroad of expressions of practices associated with urban life.[8] The study of early cinema has thus primarily focused on the modernity of the medium and seldom on the imperial forces which shaped cinema and were represented in films, and which were equally defining in my view. Hence, by bringing together the domains of early cinema and imperial history, I aim at filling a gap. In early films, imperialism was a discourse which legitimized Britain as a country whose technological superiority over colonized peoples was a key foundation of British modernity. The Victorian era was also characterized by anxiety about slum conditions and by urban degeneration which coexisted with state-managed urban development; fittingly imperial, early cinema largely focused on the latter. Film evaded crucial topics such as urban poverty and perpetuated an ideological view of reality seen through the eyes of middle-class filmmakers unconcerned with urban deprivation. Hence early films exemplify the ways in which texts can signify by what they leave out, their "structured absences", as much as by what they include.[9]

This book is preoccupied with the ways in which socially marginal groups and the spaces they inhabited were absent from cinematic representation. The work of nineteenth-century novelists like Charles Dickens and George Gissing and photographers like John Thompson, Thomas Annan and Samuel Coulthurst contributed to the creation of an idea of urban life in the slums through diverse representations and stories of the poor and the disadvantaged. In the second half of the century, social researchers like Henry Mayhew, Charles Booth and Fred Scott investigated and exposed chronic poverty and a semi-criminal underclass in Victorian society. Early films entirely neglected narratives of urban deprivation in the Victorian and Edwardian city and privileged a more prosperous cityscape made of monuments, landmarks, tourist sites, busy thoroughfares and workplaces.

8 Vanessa Schwartz, *Spectacular Realities: Early Mass Culture in* Fin-de-Siècle *Paris* (Berkeley, Los Angeles and London: University of California Press, 1998), 177–179.

9 On the concept of structured absence in semiotics, see: Keyan G. Tomaselli, *Appropriating Images: the Semiotics of Visual Representation* (Charlottesville: University of Virginia Press, 1994), 237–238.

Films of urban transport and everyday life in the suburbs also revealed the nature of urban society, and recorded the importance of self-regulation and movement in understanding urban modernity in the Age of Empire. My investigation of these films aims at contextualizing the governance of conduct as it emerged from the business of movement in the imperial city and the order within this movement. Images of disorderly movement, for example in films of popular markets, and very occasional displays of anti-imperial sentiments and resistance to the British rule in films of strikes and demonstrations will also be discussed in the following chapters.

Non-fiction films and the city in the Age of Empire

At the beginning of the twentieth century, three-quarters of the British population lived in cities and towns with over 10,000 inhabitants.[10] The nineteenth century had seen a growth in the total population from sixteen million to over forty million, as well as an increase in per capita income and other changes in the economic structure. An accelerated urbanization was driven by new forms of transport, improvements in sanitation, the concentration of commerce and industry in urban centres and an increasing division between working-class tenements in the inner city and middle-class residential suburbs.[11]

Early non-fiction films offer a great wealth of information on urban life in subsequent years. Late Victorians and Edwardians would go to film screenings to view moving images of familiar street scenes, technological novelties, and distant and exotic places. Richard C. Raack has evocatively

10 See Robert J. Morris, "Urbanization", in *Atlas of Industrializing Britain, 1780–1914*, eds John Langton and Robert J. Morris (London: Methuen, 1986), 164–167.

11 See Asa Briggs, *Victorian Cities: a Brilliant and Absorbing History of Their Development* (London: Penguin, 1990). See also: Robert J. Morris and Richard Rodger (eds), *The Victorian City: a Reader in British Urban History: 1820–1914* (London and New York: Longman, 1993).

defined each frame of an actuality film as "a separate first-hand record of the past."[12] By recording a particular area of human conduct, Raack continues, an actuality film supplies the historian with vital information not provided elsewhere.[13] "Animated photographs", as the *Daily News* put it in 1896, can be seen as the result of a journey in a time machine taking us back to a version of reality which belongs to the filmmakers and the original viewers.[14] The specific point of view of early filmmakers has been investigated by Antonio Costa, who has looked at the theme of the journey around the world in early European cinema and at the ways in which early films contributed to the creation of a new concept of space and time:

> With film, landscape, understood as the synthesis of the geographic-anthropological features of a territory, could be reproduced with an unprecedented objectivity that now included duration as well as movement. Yet, even the most neutral of shots always involves the adoption of a point of view, the presence of interpretative models and attitudes that belong to culture, ideology, and the world views of both filmmakers and viewers. Thus, landscape shown in cinema is never a pure or simple reproduction. Rather, it is a technical, economic, cultural, and semiotic (discursive) production.[15]

Early films reproduced reality but, as Costa suggests here, both the camera and the filmmakers are part of reality themselves, and cinematic reality is therefore an expression of the dominant ideology. Bill Nichols has investigated the ways in which ideology informs images in films, with a particular focus on non-fiction cinema. He writes:

12 Richard C. Raack, "Historiography as Cinematography: A Prolegomenon to Film Work for Historians", *Journal of Contemporary History*, vol. 18, no. 3 (1983): 411. This essay was written in 1983. In the following years, practitioners of film history have become more critical in the way they treat actuality films as records of the past. Raack's argument might thus appear instrumental, but it represents nevertheless a fascinating look at the undeniable visual power of early films.

13 Raack, 411–412.

14 From the *Daily News* (16 March 1896: 1). "The Cinematographe, or animated photographs as they ought to be called, could not have found a more appropriate home than the Polytechnic Institution."

15 Antonio Costa, "Landscape and Archive: Trips around the World as Early Film Topic (1896–1914)", in *Landscape and Cinema*, ed. Martin Lefebvre (New York and London: Routledge, 2006), 247.

> Ideology arises in association with processes of communication and exchange. Ideology involves the reproduction of the existing relations of production (those activities by which a society guarantees its own survival). Ideology operates as a constraint, limiting us to certain places or positions within these processes of communication and exchange. Ideology is how the existing ensemble of social relations represents itself to individuals; it is the image a society gives of itself in order to perpetuate itself.[16]

This argument can help us to understand the immediate relationship between early film and the ideological apparatus which produced it. Early actuality films should be discussed by giving priority to their social, historical and political interpretation. Only such a reading of the film text can result in a full understanding of early factual filmmaking, and its exemplification of a collective meaning of history.

As Siegfried Kracauer observed, public spaces crowded with people moving in diverse directions were filmed by pioneer filmmakers exploiting the ability of the new medium to record and reveal reality.[17] In *One-Way Street*, Walter Benjamin also explained how activities such as driving through busy thoroughfares can only be visually recast in films. He suggested that the moving picture "commands optical approaches to the essence of the city."[18] In 1928, Benjamin was the first theorist to discuss the privileged relationship between cinema and urban space that began with the invention of cinema. As the Lumière brothers' early views of Lyon and Paris or the moving pictures of New York produced by Thomas Edison demonstrated in the 1890s, the built environment and film established an axiomatic relationship from the first days of the cinematograph. As Tom Gunning suggests:

16 Bill Nichols, *Ideology and the Image* (Bloomington: Indiana University Press, 1981), 1.

17 Siegfried Kracauer, *Theory of Film: the Redemption of Physical Reality* (Princeton, NJ: Princeton University Press, 1997; first published in 1960, London and New York: Oxford University Press), 31–33.

18 Walter Benjamin, *One-Way Street and Other Writings* (London and New York: Verso, 1985; *Einbahnstraße*, originally published in 1928 in Berlin), 298.

> The first film shows were primarily big city affairs. Nearly all early film documents present a *mise en abîme* of audiences filling vaudeville halls from busy city streets in order to see projected on the screen… busy city streets. The transfer to film allowed the city street to become another sort of spectacle, one mediated by an apparatus and filled with endless attraction.[19]

Giuliana Bruno has also emphasized the relationship between early cinema and the modern street, arguing that early films provided new spectatorial means of transportation through the streets of the city "as an intentional phenomenon, insistently portraying urban space by reproducing the captivating fluvial motion of cities."[20] The study of British cinema, however, can challenge the assumption that visual attraction of early films always sought to shock the audience. According to Gunning:

> Theatrical display dominates over narrative absorption, emphasizing the direct stimulation of shock or surprise at the expense of unfolding a story or creating a diegetic universe.[21]

Gunning is here referring to fiction films produced in the United States and, in particular, to the films made by American film pioneer D.W. Griffith. Similarly, Charles Musser has investigated the origins of motion picture technologies and the social and economic aspects of the beginnings of American cinema. Like Gunning, Musser traces the development of an early

19 Tom Gunning, "Images of the City in Early Cinema", unpublished paper presented at the Getty Centre, *Cinecity Film and Perceptions. of Urban Space*, Santa Monica, California, 23 March 1994, reported in Helmut Weihsmann, "The City in Twilight: Charting the Genre of the City Film 1900–1930", 9. See also Tom Gunning, "From the Kaleidoscope to the X-Ray: Urban Spectatorship, Poe, Benjamin, and *Traffic in Souls*", *Wide Angle*, vol.19, no. 4 (1997): 25–61.

20 Giuliana Bruno, *Atlas of Emotions, Journeys in Art, Architecture, and Film* (London and New York: Verso, 2002) 183–184. See also Giuliana Bruno, "Site-seeing: Architecture and the Moving Image", *Wide Angle*, vol. 19, no. 4 (1997): 8–24.

21 Tom Gunning, "Cinema of Attractions: Early Cinema, Its Spectator and the Avant-Garde", in *Early Cinema: Space Frame Narrative*, ed. Thomas Elsaesser (London: BFI Publishing, 1990), 62. See also Tom Gunning, *D.W. Griffith and the Origins of American Narrative Film: the Early Years at Biograph* (University of Illinois Press: Urbana, 1994).

cinema of attractions into a cinema of narrative complexity.[22] That early cinema was an attraction which aimed at causing shock in the spectator is one of the most widely held ideas used in the study of films made in the first decade of cinema. When we take into account the specificity of place and its distinct ensemble of wider social relations, however, this model does not fit neatly the themes and aesthetics of early films of the city. In particular, empire was a local, national and global force which played a dominant role in early British cinema. British films require a somewhat different model to that suggested by Gunning and Musser. This book thus investigates the application of cinematographic technologies to the ability to situate oneself in the empire as it emerges from the relationship between non-fiction films and geographical imagination. This dimension is more often than not absent in studies of early cinema, which tend to favour a focus on the modernity of the medium and often pay insufficient attention to the specificity of the place where the films were made and its social and historical implications.

My investigation will begin with the specific case of London in early films, and thus addresses what Gunning – embarking from Edgar Allan Poe's short story *The Man of the Crowd* – describes as "the energy and vitality of urban bustle" in London.[23] Gunning, who has investigated elements of urban spectatorship in relation to urban modernity, has characterized the early cinematic city as a "constantly varied, kaleidoscopic spectacle."[24] Indeed, scholars like Gunning and Musser have built a very successful framework of analysis for early films of cities such as New York, but one that can be partly limited when applied to other national contexts. In this book I shall still take into account the modernity of the city and its cinematic expressions; urban bustle does emerge from films of London, Manchester or Liverpool. However, an entirely different dimension is given by the orderly movement of imperial processions and parades, and my main focus will be

22 Charles Musser, *The Emergence of Cinema: the American Screen to 1907* (University of California Press: Berkeley and Los Angeles, 1994).

23 Gunning, "From the Kaleidoscope to the X-Ray: Urban Spectatorship, Poe, Benjamin, and *Traffic in Souls*", 25.

24 Gunning, "From the Kaleidoscope to the X-Ray: Urban Spectatorship, Poe, Benjamin, and *Traffic in Souls*", 42.

on the ways in which empire provided a narrative into which films fitted coherently and defined the city in ways in which are not accounted for by Gunning and Musser. Gunning has focused on the formal development of filmic discourse and has argued that the so-called cinema of the attractions was replaced by narrative cinema.[25] Based on spectacle and the thrills and shocks created by the moving pictures in its early days, cinema developed into a narrative form defined by a more complex use of editing. My aim is to demonstrate that in the case of early British cinema, imperialism offered a pre-existing narrative by connecting places, people and time in the cinematic space of the British Empire. Filmmakers did so by employing and reimagining a series of preceding discourses emerging from other media, including magic lantern shows, photography and illustrated magazines. Hence, early British films are the result of a tension between modernity and tradition. British imperialism was a modern phenomenon, at once both the engine and the outcome of the Industrial Revolution. And yet, British imperial identity coexisted with the weight of old tradition embodied in the monarchy; the relationship between present and past thus resulted in a dialectic process of change and continuity. For example, the reinvention of the Iron Age warrior queen Boudica as an icon of female leadership and strength, as Vanessa Collingridge suggests, became a reference point for Queen Victoria and a symbol of empire.[26] Once Victoria became empress of India in 1877 all royal occasions were also imperial occasions and the late nineteenth century became the age of the invented traditions, as it is suggested by David Cannadine.[27] This tension between modernity and tradition is expressed by early moving images of British and colonial cities. In particular, films of urban parades and processions result in a significant tension between the old and the new. On the one hand, these events reflected

25 See, for example: Gunning, *D.W. Griffith and the Origins of American Narrative Film: the Early Years at Biograph* (Urbana: University of Illinois Press, 1991).

26 Vanessa Collingridge, *Boudica: the Life of Britain's Legendary Warrior Queen* (New York: Overlook Press, 2007), 355.

27 David Cannadine, "The Context, Performance and Meaning of Ritual: the British Monarchy and the *Invention of Tradition*, c. 1820–1877", in *The Invention of Tradition*, eds Eric Hobsbawm and Terrence Ranger (New York and Cambridge: Cambridge University Press, 2000), 124.

the opulence of a modern empire defined by unprecedented global expansion and control. On the other hand, the image of these rituals is one of heraldry and tradition. The regular movement of soldiers and horses on the streets of the city suggests a conflict with that fragmented experience of urban modernity which characterizes other kinds of street scenes both in Europe and in the United States. Overall, as I shall try to demonstrate in the following chapters, these images reveal a coherent cinematic narrative preceding the birth of narrative cinema in the canonical sense.

Imperialism can be understood as an alternative framework for the study of early films and simultaneously as an instrument to reassess modernity as a vital idea for the understanding of the early days of cinematography. On these basis, the present investigation addresses imperialism as the image late Victorian and Edwardian Britain gave of itself in order to perpetuate itself. Imperialism, as W.J.T. Mitchell puts it, is "a complex system of cultural, political, and economic expansion and domination that varies with the specificity of places, peoples, and historical moments."[28] Its relation to landscape, or cityscape, Mitchell suggests, is offered as an answer to an inquiry on a space disclosing both "utopian fantasies of the perfect imperial prospect and fractured images of unresolved ambivalence and unsuppressed resistance."[29] The aim of this book is to understand the ways in which early films of the city reflected the duality of imperial landscape and were themselves fantasies of perfect imperial prospect. The early years of cinematography coincided with the apogee of the imperialist ordering of the globe. As Ella Shohat suggests, the twin beginnings of cinema and psychoanalysis, cinema and nationalism, cinema and consumerism have all been explored in depth but, on the other hand, the coincidence between the beginning of cinema and the heights of the imperial project has not been fully explored.[30] This study aims at reassessing this relationship.

28 William J.T. Mitchell, "Imperial Landscape", in *Landscape and Power*, eds William J.T. Mitchell (Chicago and London: University of Chicago Press, 2002; first published in 1994), 9.

29 Mitchell, 10.

30 Ella Shohat, *Unthinking Eurocentrism: Multiculturalism and the Media* (London and New York: Routledge, 1994), 100–102.

Interdisciplinary threshold: Films and urban space

The history of the cinematic city begins in Bridge End, Leeds, on a bright and sunny day in October 1888. On that day Louis Le Prince, a Frenchman from Metz in his mid-forties, entered the building then occupied by the Hicks Brothers Ironmongers Company and, from the second storey window of the building at the southeast corner of the bridge, filmed the earliest moving pictures of a city. Bridge End was one of the busiest thoroughfares in Leeds, and Le Prince's single-lens motion picture camera captured movement in the city at sixteen frames per second: horse-drawn coaches and carts are visible on the two-ways street and a number of pedestrians can be seen crossing the street or walking along it. Two years later, on 16 September 1890, the unsung inventor of the cinematograph boarded the 2.42 pm train from Dijon to Paris and disappeared on the eve of the first public demonstration of his work: his body and luggage were never found.[31] The history of the cinematic city was resumed in March 1895, when the Lumière brothers held their first private screening of moving pictures.

As Stephen Barber puts it, Le Prince's *Leeds Bridge* initiated the capacity of film "to transfuse to its spectators a compacted charge of nostalgia for the momentary apparitions of cities or human bodies whose forms were just about to disappear forever."[32] More broadly, in *Projected Cities: Cinema and Urban Space*, Barber has investigated the relationship between cinema and the city, and the spatial power of film in European and Japanese cinema from the origins to the digital revolution. The first part of Barber's book is concerned with the cinematic city in silent films and presents examples from the films of Le Prince, the Skladanowsky Brothers and Dziga Vertov: cinema, Barber argues, was born as an urban phenomenon both in terms of subjects and audiences. Just as a city is a single variant of the process of

31 Recently, it has been claimed that a photograph of a drowned man in the Paris police archives is that of Louis Le Prince.

32 Stephen Barber, *Projected Cities: Cinema and Urban Space* (London: Reaktion Books, 2002), 19.

urban transformation, Barber continued, cinema is a variant of the multiple history of the potential image.[33]

Barber is one of several scholars who have focused their studies on the previously neglected relationship between cinema and the city. This shift has followed the work of a number of cultural theorists and sociologists who, since the 1970s, have focused on space, including Henri Lefebvre, Michel Foucault, Marshall Berman, David Harvey, Fredric Jameson, Edward Soja, and Mike Davis.[34] As Mark Shiel suggests:

> The "spatial turn" in social and cultural theory on the Left (broadly defined) since the 1970s [...] has involved a growing recognition of the usefulness of space as an organizing category, and the concept of "spatialization" as a term for the analysis and description of modern, and (even more so) of postmodern, society and culture.[35]

Shiel explains that one of the premises of the study of the cinematic city is that interdisciplinary contact between Film Studies, Sociology, Geography, Architecture, and other disciplines, can help in addressing issues shared by these disciplines because of the close relationship in cinema between culture, society and space:

> Thematically, the cinema has, since its inception, been constantly fascinated with the representation of distinctive spaces, lifestyles, and human conditions of the city [...]. Formally, the cinema has long had a striking and distinctive ability to capture and

33 Barber, 15.

34 See: Lefebvre, *The Production of Space*; Michel Foucault, *Discipline and Punish* (London: Allen Lane, 1977); Marshall Berman, *All That is Solid Melts Into Air: The Experience of Modernity* (New York and London: Penguin, 1988; first published by Simon & Schuster, 1982); David Harvey, *The Condition of Postmodernity: An Enquiry into the Origins of Cultural Change* (Oxford and New York: Basil Blackwell, 1989); Fredric Jameson, *Postmodernism, or, the Cultural Logic of Late Capitalism* (London: Verso, 1991); Edward W. Soja, *Postmodern Geographies: The Reassertion of Space in Critical Social Theory* (London and New York: Verso, 1989); Mike Davis, *City of Quartz: Excavating the Future in Los Angeles* (London: Verso, 1990).

35 Mark Shiel, "Cinema and the City in History and Theory", in *Cinema and the City: Film and Urban Societies in a Global Context*, eds Mark Shiel and Tony Fitzmaurice (Oxford and New York: Blackwell Publishing, 2001), 5.

> express the spatial complexity, diversity, and social dynamism of the city through *mise-en-scène*, location filming, lighting, cinematography, and editing [...].[36]

Anthony Sutcliffe's work on the metropolis in cinema, which focuses on a range of international film examples from the 1920s and 1930s, was published in 1984 in an anthology examining perceptions of the city during the period 1890–1940, and is one of the earliest examples of a study of the cinematic city.[37] From the early 1990s, there has been an increasing concentration of monographs and edited books from different academic backgrounds about the relationship between films and cities. Bruno's *Streetwalking on a Ruined Map*, a study of the films of Elvira Notari and Naples, was published in 1992 and became one of the most influential studies of the cinematic city.[38] Two years later, Christine Boyer investigated the ways in which cinema, among a broader series of different visual and mental models, has recognized, depicted, and contributed to shaping the urban environment. Boyer also argued that great cities at the turn of the twentieth century seized the centre of cinema's public stage, and film deflected the incongruous visual data offered by urban life into new meanings and forms.[39] In 1997, David Clarke edited *The Cinematic City*, an anthology of essays devoted to a diverse selection of films and aimed at exploring the ways in which the city and cinema have shaped one another.[40] Clarke argued that the relationship between cinema and the city had been broadly neglected in both film theory and urban theory, and that the implicit centrality of the city to film has worked against an explicit consideration of its importance.[41] His book explores the cinematic city without impos-

36 Shiel, 1.

37 Anthony Sutcliffe, *Metropolis 1810 –1940* (London: Mansell Publishing, 1984), 147–171.

38 Giuliana Bruno, *Streetwalking on a Ruined Map: Cultural Theory and the City Film of Elvira Notari* (New Jersey: Princeton University Press), 1992.

39 Christine Boyer, *The City of Collective Memory: Its Historical Imagery and Achitectural Entertainments* (London and Cambridge, Massachusetts: MIT Press, 2001; first published in 1994).

40 David Clarke, *The Cinematic City* (New York and London: Routledge, 1997).

41 Clarke, 1–2.

ing a single interpretation, and puts forward diverse approaches ranging from the engagements with particular films to broader discussions on the cultural meaning of the cinematic city.

In *Imagining the Modern City* (1999), James Donald studied the world cities which have inspired representational processes and discussed the ways in which artists, writers and filmmakers have shaped the experience of the city through representation.[42] Donald looked at late nineteenth century and twentieth century cityscapes in novels, paintings, photography and films. The third chapter of *Imagining the Modern City*, "Light in Dark Spaces", is devoted to the study of the cinematic city or to that archive of urban images offered by, as Donald put it, "the cities conjured up in cinema." Donald looked at the city as "a historically specific mode of seeing" whose bond with cinema slips in human culture and society at large. Cinema, Donald suggested, has always sought to teach ways of seeing and imagining the city as a sensory and symbolic experience at the same time.[43]

The interdisciplinary approach to the study of the cinematic city has also brought together Film Studies and Architecture. In 1992, Anthony Vidler intervened in a debate surrounding the built environment, and explored the role of architecture in the construction of sets and the ability of film to construct its own architecture. In particular, he studied the sets of German expressionist films such as Robert Wiene's *The Cabinet of Dr Caligari* (1920) and Paul Wegener's *Der Golem* (1920).[44] Edited by François Penz and Maureen Thomas in 1997, the volume *Cinema and Architecture* was concerned with the representation of the real and the ideal in the urban environment.[45] Helmut Weihsmann's chapter – "The City in Twilight: Charting the Genre of the City Film 1900–1930" – focused on

42 James Donald, *Imagining the Modern City* (London: Athlone Press, 1999).

43 Donald, 63, 92.

44 Anthony Vidler, "The Explosion of Space: Architecture and the Filmic Imaginary", *Assemblage: A Critical Journal of Architecture and Design Culture*, no. 21 (1993): 45–59, and *The Architectural Uncanny: Essays in the Modern Unhomely* (Cambridge, Massachusetts, London, England: MIT Press, 1992).

45 François Penz and Maureen Thomas (eds), *Cinema and Architecture* (London: BFI Publishing, 1997).

the image of the metropolis in early cinematography and argued that film is analogous to the modern perception of the city, with its "continuous sequences of space frames perceived through time."[46] Weihsmann took into account a number of study cases, from the Scandinavian films directed by Urban Gad and Victor Sjöström to Fritz Lang's *Metropolis* (1927) and László Moholo-Nagy's *Großstadt-Zigeuner* (1932). Like Anthony Vidler and Giuliana Bruno before him, Weihsmann was one of relatively few scholars who have focused on the city in silent cinema.[47] Recently, Jon Burrows has investigated the representation of the London East End district of Limehouse in silent films, and drawn attention to the specific historic and cinematic determinations of the area.[48] The complex relationships between London and its cinematic geography is also the subject of a book written by Charlotte Brunsdon. Her work looks at the cinematic city in the second half of the twentieth century. Brunsdon investigates recurring patterns of representation associated with films of the city, and thus addresses the way in which London-set films shaped our broader idea of the cinematic city.[49]

These diverse interdisciplinary studies have privileged fiction films and focused on a wide variety of case studies, in terms of the type of films, historical eras, and sites examined. Each brings together the study of films and Urban Sociology, Cultural Geography or Architecture. This book aims at applying this interdisciplinary framework to the specific context of the British imperial and colonial city at the end of the long nineteenth

46 Weihsmann, "The City in Twilight: Charting the Genre of the City Film 1900–1930", in *Cinema and Architecture*, 9.

47 Marco Bertozzi, who studied the city in early Italian films, and John Fullerton, who investigated the image of Stockholm in early actuality films, also contributed to this field. See John Fullerton, "Seeing the World with Different Eyes, or Seeing Differently", in *Nordic Explorations: Films Before 1930*, eds John Fullerton and Jan Olsson (Sydney, John Libbey & Company Pty Ltd, 1999), 163–186; Marco Bertozzi, "Visualising the Past: the Italian City in Early Cinema", *Film History*, vol. 12, no. 3 (2000): 322–329.

48 Jon Burrows, "A Vague Chinese Quarter Elsewhere: Limehouse in the Cinema, 1914–36", Journal of British Cinema and Television, vol. 6 (2009): 282–301.

49 Charlotte Brunsdon, *London in Cinema: the Cinematic City Since 1945* (London: BFI Publishing, 2008).

century.[50] Most studies of early British cinema have not only neglected urban space as a discursive category but also largely ignored non-fiction filmmaking. A clear distinction between fiction and non-fiction films in the early cinema can at times be problematic, with a thin line between the two. However, the focus on urban space can facilitate such categorization; non-fiction films, including street scenes or processions, were filmed on location. Fiction films, or films with hybrid elements of fiction and non-fiction forms, would normally avoid urban locations, often unfeasible for filming this type of narratives. Well into the 1910s filmmakers such as Bert Haldane will film on location in the streets of London, with films such as *The Lure of London* (1914), *The Rogues of London* (1915), and *As a Man Sows: or, an Angel of the Slums* (1914); in the early days of cinematography, however, urban life was thus primarily deflected in cinematic form through actuality films.

A long-standing neglect: Early non-fiction filmmaking

At the end of the nineteenth century the great majority of films produced in Britain were non-fiction, and during the first decade of the twentieth century the output of fiction and non-fiction films was broadly equal. Technological limitations, the novelty of the medium and its connections with photography, as Vanessa Schwartz suggests, made of non-fiction cinema the main type of film production in the early days of cinematography.[51] Non-fiction cinema was influenced by pre-existing representations of reality which culminated with the commercial success of photographic reproduction as the privileged technique to capture the real. Early actuality

50 The long nineteenth century refers to the period between the years 1789 and 1914. See Hobsbawm's work: *The Age of Revolution: Europe 1789–1848*, *The Age of Capital: Europe 1848–1875*, and *The Age of Empire: Europe 1875–1914*.

51 Schwartz, 190–193.

films were born of the same realist impulse and provided a vivid record of their time. Actuality, topical, interest and travel films covered a wide range of subjects, such as royal, military and state events, sports, and exhibitions, as well as ordinary events and everyday life.[52] The pioneering nature of early filmmaking makes it difficult to give precise statistics on early film production. According to surviving records, over 5,300 non-fiction films were produced in Britain in the period 1895–1914, and given the loss of several records, the total sum is likely to be much higher than this. Between 1895 and 1900 the production of non-fiction film grew from 25 to 450 films a year and the number of fiction films from 4 to just over 100 films a year. From the beginning of the twentieth century an average of 250 fiction films and the same number of non-fiction films were produced in Britain each year. Only in 1912 did the production of fiction films begin to outnumber that of non-fiction films with almost 600 fiction films against just over 200 non-fiction films.[53]

Non-fiction films tackled a great number of themes and were mostly shot in cities because of the attractions offered by their busy streets and also because of the urban location of most film companies.[54] However, notwithstanding their greater number, the study of non-fiction British films produced before the Great War has largely been neglected in favour of fiction films from the same period. With some relevant exceptions, most histories of early cinema in Britain have focused on the fiction work of filmmakers such as Robert W. Paul and Cecil Hepworth, and the Brighton

52 See Richard M. Barsan, *Non-Fiction Film: a Critical History* (Bloomington, IN: Indiana University Press, 1992).

53 These statistics are based on surviving records reported in the work of Denis Gifford: *The British Film Catalogue: Fiction Film, 1888–1994* (London and Chicago: Fitzroy Dearborn, 2000; 3rd edn); *The British Film Catalogue: Non-Fiction Film, 1888–1994* (London: Fitzroy Dearborn, 2001; 3rd edn).

54 Fiction films such as comedies and trick films were rarely filmed in urban locations until the mid-1910s, when feature length films became the main type of cinematographic production in the country (see, for example, Bert Haldane's *The Lure of London*, 1914, *The Lights of London*, 1915, *The Rogues of London*, 1915, and *Darkest London*, 1915).

pioneers George Albert Smith and James Williamson.[55] This is also true of the first comprehensive account of early British cinema by the French film historian Georges Sadoul in 1945, which argued that editing was discovered in England by some filmmakers of the so-called Brighton School, and offered an account of the use of editing in a number of films directed by Smith, such as *Grandma's Reading Glass* (1900), and Williamson, such as *Attack on a China Mission* (1900).[56]

In the late 1940s, Rachael Low published *The History of the British Film: 1896–1906* and *The History of the British Film: 1906–1914*, the first two volumes of a series on British cinema; this work was based on research by the History Committee of the British Film Institute, whose chairman was Cecil Hepworth himself.[57] Low gave a full account of film production, distribution and exhibition in the period 1896–1914 and described the types of films produced in Britain during those years, including non-fiction films. Low was thus the first scholar to offer a discussion of non-fiction films from the early days of cinematography in Britain. Published over thirty years later, Michael Chanan's *The Dream that Kicks* (1980) tackled the work of the British pioneers and discussed the popular and visual influences on the culture of early cinema, i.e. the magic lantern, photography, itinerant

55 See, for example, Charles Oakley, *Where We Came In: Seventy Years of the British Film Industry* (London: George Allen and Unwin, 1964); George Perry, *The Great British Picture Show* (London: Pavilion Books, 1985; first published in 1974); Roy Armes, *A Critical History of the British Cinema* (New York: Oxford University Press, 1978); David Shipman, *The Story of Cinema: from the Beginning to* Gone with the Wind (London, Sydney, Auckland, Toronto: Hodder and Stoughton, 1982); David Bordwell, and Kristin Thompson, *Film History: An Introduction* (New York: McGraw-Hill, 1994).

56 George Sadoul, "Early Film Production in England: the Origin of Montage, Close-Ups, and Chase Sequence", *Hollywood Quaterly*, vol. 1, no. 3 (1946): 249–259. This essay was translated by Y. Templin and originally published by *Cinéma* in 1945. See also: George Sadoul, *Histoire Générale du cinema: 2. Les pionniers du cinéma 1897–1909*, Paris: Denoël, 1948.

57 Rachael Low, *The History of the British Film: 1896–1906* (London: Gerges Allen & Unwin, 1948); Rachael Low, *The History of the British Film: 1906–1914* (London: Gerges Allen & Unwin, 1949).

entertainers and the music hall.[58] Chanan explored the broader context given by the dialectic of the invention of film, the cultural and economic context and the industrial growth of the film market. By dealing with the economics of the film business, the audience reception of early cinema and the formal development of cinema between 1895 and 1910, Chanan explained that early films can be seen as a cultural commodity because the new-born cinema was a product of nineteenth-century capitalism. This argument revealed cultural and ideological implications absent in previous works on early British cinema. Noël Burch also explored early British cinema and its cultural and economic context. In *Life to Those Shadows* (1990), Burch criticized the assumption that film language is a natural development which transcends history and cultures and, as Chanan did, argued that it has a social and economic history, which is determined by the time and the place where cinema was born, i.e. the capitalist and imperialist West at the turn of the twentieth century. By examining the visual devices employed by the filmmakers and internalized by the audience, Burch explored the socio-historical circumstances which led to the institutionalization of cinematic language.[59] Both Chanan and Burch touched briefly upon non-fiction filmmaking, although their arguments in favour of a social and economic history of the language of film and the film medium can effectively be applied to non-fiction filmmaking and will be taken into account throughout this book.

During the past twenty years revisionist studies of early film history and a radical rethinking of British cinema have often intersected. Justine Ashby and Andrew Higson, for example, have published original contributions to the study of British cinema.[60] Notwithstanding the publication of chapters on non-fiction filmmaking, the result of these studies has not

58 Michael Chanan, *The Dream that Kicks: the Prehistory and Early Years of Cinema in Britain* (London and New York: Routledge, 1996; first published in 1980).

59 Burch, 80–108.

60 Revisionist approaches to the national cinema have been a vital part of recent film studies, and include books such as: Andrew Higson ed., *Dissolving Views: Key Writings on British Cinema* (London: Cassell, 1996); Andrew Higson, *Waving the Flag: Constructing a National Cinema in Britain* (Oxford: Clarendon Press, 1995);

led to an in-depth re-evaluation of early non-fiction filmmaking in Britain. *Re-Viewing British Cinema*, edited by Wheeler Winston Dixon in 1994, presents one significant chapter on non-fiction filmmaking, an exception in a field dominated by the study of fiction films.[61] Stephen Bottomore's "Frederic Villiers: War Correspondent" is a detailed study of the first British war cinematographer whose non-fiction films and contribution to national cinema had hardly been assessed previously.[62] Bottomore is one of the few scholars concerned with early non-fiction films, and he has identified the greater popularity enjoyed by fiction films since then as the reason for the long-standing neglect of early and silent factual film.[63]

In 1994 a significant contribution to the study of early non-fiction cinema was made by the discovery in a cellar in Blackburn of eight hundred actuality films produced during the first ten years of the twentieth century by the Sagar Mitchell and James Kenyon Company. These films were later restored at the National Film and Television Archive, broadcasted on the BBC and released on DVD. These actuality films were also the subject of an anthology including chapters by leading scholars from a variety of fields and of a monograph by Vanessa Toulmin.[64] The latter, *Electric Edwardians: The Story of the Mitchell & Kenyon Collection*, contains a long chapter on the Edwardian city and the ways in which the Mitchell and Kenyon films captured the movement of the streets and public places

---

Justine Ashby and Andrew Higson (eds), *British Cinema, Past and Present* (London and New York: Routledge, 2000).

61 Wheeler W. Dixon ed., *Re-Viewing British Cinema, 1900–1992: Essays and Interviews* (New York: State University of New York Press, 1994).

62 Stephen Bottomore, "Frederic Villiers: War Correspondent", in *Re-Viewing British Cinema, 1900–1992: Essays and Interviews*, 11–23.

63 See Stephen Bottomore, "Rediscovering early non-fiction film", *Film History*, vol. 13 (2001), 160–173; Stephen Bottomore, "Have You Seen the Gaekwar Bob?: Filming the 1911 Delhi Durbar", *Historical Journal of Film, Radio and Television*, vol. 17, no. 3 (1997), 309–345; Stephen Bottomore "An Amazing Quarter Mile of Moving Gols, Gems and Genealogy: Filming India's 1902/03 Delhi Durbar", *Historical Journal of Film, Radio and Television*, vol. 15, no. 4 (1995), 495–515.

64 Vanessa Toulmin, Simon Popple and Patrick Russell, *The Lost World of Mitchell and Kenyon: Edwardian Britain on Film* (London: British Film Institute, 2004).

through a series of street scenes and phantom journeys filmed from the front of trams and trains.[65] Films and transport in the Edwardian city are also the themes of two essays by Patrick Keiller and Ian Yearsley included in *The Lost World of Mitchell and Kenyon: Edwardian Britain on Film.*[66] Keiller has also looked at early films and urban space in two recent essays focusing on Kennedy-Laurie Dickson's *Panorama of Ealing from a Moving Tram* (1901).[67] Lynda Nead's work on the urban rhythm of London in early non-fiction cinema is also part of a new recognition of the cinematic city as subject of investigation.[68] The research of Toulmin, Keiller and Nead has contributed to a better understanding of the ways in which city life was deflected into the film form and has informed my own view on the subject. This book draws upon their work and aims at developing a more extensive narrative of urban, imperial and filmic cultures. In order to do this, I should now contextualize early cinema as part of an urban phenomenon in middle-class hands.

65 Vanessa Toulmin, *Electric Edwardians: the Story of the Mitchell & Kenyon Collection* (London: BFI Publishing, 2006), 154–201.

66 Patrick Keiller, "Tram Rides and Other Virtual Landscapes", in *The Lost World of Mitchell and Kenyon: Edwardian Britain on Film*, 191–2000; Ian Yearsley, "On the Move in the Streets: Transport Films and the Mitchell and Kenyon Collection", in *The Lost World of Mitchell and Kenyon: Edwardian Britain on Film*, 181–190.

67 See Patrick Keiller, "Urban space and early cinema", in *Cities in Transition: the Moving Image and the Modern Metropolis*, eds Andrew Weber and Emma Wilson (London: Wallflower Press, 2008), 29–39; Patrick Keiller, "The City of the Future", in Laraine Porter and Bryony Dixon, *Picture Perfect: Landscape, Place and Travel in British Cinema before 1930* (Exeter: the Exeter Press, 2007), 104–112.

68 Lynda Nead, "Animating the Everyday: London on Camera c.1900", *Journal of British Studies*, vol. 43, no. 1 (2004): 65–90; Lynda Nead, *The Haunted Gallery Painting, Photography and Film around 1900* (London: Yale University Press, 2007), 107–131.

Early cinema and urban space: Middle-class filmmaking in context

At the turn of the twentieth century, film production in Britain was among the most advanced in the world. Both Roy Armes and Noël Burch have investigated the social and professional origins of pioneer filmmakers and defined early film production in Britain as "middle-class cinema."[69] Burch has argued that the development of popular leisure in Britain in the nineteenth century led to the identification of the new-born cinematograph as a popular spectacle aimed at working-class people but in middle-class hands.[70] From its outset, British cinema was in the hands of middle-class Victorian entrepreneurs, such as Cecil Hepworth, Robert W. Paul, George A. Smith, James Kenyon and Sagar Mitchell, who had previously worked as photographers, lanternists, or producers of photographic material and lantern slides. Hepworth was one of the few pioneers who continued making films after the Great War, while Paul and many others returned to instrument making, photography and similar professions. Charles Urban, who started making films in 1898, was later employed by the British Topical Committee for War Films to promote the British war effort during the Great War.

Characters and edifying story-lines in films made in the period 1895–1914 often revealed the middle-class origins of the filmmakers, which were also reflected in non-fiction films revealing acceptance of, and often enthusiasm, for the imperial adventure. According to Richard N. Price, for the lower middle class, imperialism was an assertion of responsibility made attractive by the tensions that existed within Victorian and Edwardian societies. In the love of country and empire, middle-class subjects thus found an answer to the anxieties related to social and economic status which were caused by the fear of being subject to the erratic and unpredictable forces

69 Roy Armes, *A Critical History of the British Cinema* (New York: Oxford University Press, 1978) 17; Noël Burch, *Life to Those Shadows* (London: BFI Publishing, 1990), 92.

70 Burch, 80–108.

of the market.[71] Indeed, the middle class struggled to consolidate its position against threats to its own and its empire's survival. Middle-class fears were generated by the possibility of insurrections suggested by urban riots, demonstrations and rallies. The urban mob, personified by the hooligan, posed a threat to the middle class, and gave a physical dimension to the widespread fear of political turmoil and social unrest; this fear prompted theories of urban decay and suggestions that national life was menaced by diverse threats.[72] International threats such as the expansion of German maritime power and the industrial competition from America menaced the British middle class from the mid 1900s. Internal threats appeared in the form of organized labour and the growing influence of socialism which menaced the middle class from the mid-1880s. An answer to these threats was given by the growing attempt by the middle class and the aristocracy to establish the hegemony of the imperial ideology. This attempt used education, recreation and popular entertainment which, from the mid-1890s, included cinematography. Middle-class filmmakers were thus drawn to displays of imperial power. They generated an image of the city where imperial landmarks and events were part of an orderly urban space characterized by the regular movement of troops parading on the streets, the smooth movement of the newest forms of urban transport, and by economic dynamism and cultural displays expressing imperial order. The ideological weight of early films, arguably more consistent than in other forms of popular culture, seemed to testify the presence of an increased sense of instability and of a plurality of treats. The fears of the urban middle class were thus exorcized in a cinematic image which was the result of an ideological process of selection of the themes and subjects to portray on film. As Lynda Nead suggests, the imperial project depended on the eradication of disorder and the regulation of the unruly spaces of the city and on the

71 Richard N. Price, "Society, Status and Jingoism: the Social Roots of Lower Middle Class Patriotism, 1870–1900", in *The Lower Middle Class in Britain 1870–1914*, ed. Geoffrey Crossick (London: Croom Helm, 1977), 89–112.

72 See Steve Attridge, *Nationalism, Imperialism and Identity in Late Victorian Culture: Civil and Military Worlds* (New York: Palgrave Macmillan, 2003), 107–108.

triumph of the modern over outmoded forms of society.[73] Cinematic cities were essentially free from disorder and dissent, and thus a projection of the imperial rule. Common themes and the specificities of place thus emerged from a middle-class cinema generating an urban spectacle of modernity and empire aimed at working-class and lower middle-class audiences.

Imperialism, as Hobsbawm suggests, dramatized the triumph of the middle classes of the Western countries and the colonial societies created in their image.[74] Films became a means to express this triumph, and revealed a world created in the image of the middle class to which the filmmaker belonged. As John MacKenzie has argued, early filmmakers recognized the propaganda potential of cinematography at once and, by producing anti-Boer propaganda or films attacking Jewish immigration, their work often reflected the aggressive militarism of juvenile literature and the triumphs of the West over the East.[75] Celluloid propaganda developed an obsession with war and warfare: R.W. Paul and Cecil Hepworth produced in 1900 several films on the army and the navy, which continued the nineteenth-century tradition of music hall military drama. An example of the conservative ideology of pioneer filmmakers is provided by Hepworth's *The Aliens' Invasion* (1905), which was openly against the arrival in Britain of foreign workers.[76] This film, which is now lost, denounced the immigrant's customs and was in tune with the nationalistic feelings expressed, for example, on the pages of the *Daily Mail.* On 3 February 1900, a journalist reported a doubtful reconstruction on the arrival of 600 refugees at Southampton:

> There were Russian Jews, Polish Jews, German Jews, and Peruvian Jews: all kinds of Jews, all manner of Jews. They fought and jostled for the foremost places at the gangways; they rushed and pushed and struggled into the troops shed, where the Mayor of Southampton had provided free refreshments. They had breakfasted well on board, but they rushed as though starving at the food. They brushed the attendant

73 Lynda Nead, *Victoria Babylon: People, Streets and Images in Nineteenth-Century London* (New Haven, CT and London: Yale University Press, 2000), 194–195.

74 Hobsbawm, *Age of Empire*, 81.

75 John M. MacKenzie, *Propaganda and Empire* (Manchester and New York: Manchester University Press, 1986), 34–36.

76 Burch, 107.

> to one side, they cursed if they were not quickly served, they helped themselves at will, they thrust the children to the background, they pushed the women, they jostled and upset the weak, they spilled coffee on the ground in wanton waste.[77]

The *Daily Mail* aimed at the newly literate lower middle-class market.[78] Films aimed at a larger audience, including illiterate working classes, and provided an ideological version of reality seen through the eyes of middle-class entrepreneurs who saw in the new-born cinematograph a profitable innovation able to support the imperial order in places of popular recreation. Until the growth of picture palaces in 1909, films were screened in music halls, fairs, and town halls.[79] The music hall, arguably the main focus of commercial attention for early filmmakers in Britain, was the dominant form of popular entertainment in nineteenth-century Britain and its shows were aimed at working-class and lower middle-class audiences. Painted by impressionist artist W.R. Sickert in the 1880s as examples of scenes of working-class leisure, in 1901 the music hall was condemned by John A. Hobson for promoting exploitative imperialist policies, the Boer War and jingoism.[80] The display of patriotic, militarist and imperial sentiments within music hall entertainment was indeed part of shows which also included the non-political acts of comedians like Dan Leno and a young Charlie Chaplin, and singers like Daisy Wood and Ella Shields. By focusing on music halls as agents of propaganda, historians and cultural theorists have elucidated the complexities of the social and cultural experience of the music hall.[81] At the turn of the twentieth century music hall acts

77 "So-Called Refugees", *Daily Mail*, 3 February 1900.

78 Paul Manning, *News and News Sources: a Critical Introduction* (New York: SAGE, 2001), 83.

79 On film exhibition in London see: Jon Burrows, "Penny Pleasures: Film Exhibition in London During the Nickelodeon Era, 1906–1914", *Film History*, vol. 16, no. 1 (2004): 60–91, and "Penny Pleasures II: Indecency, Anarchy and Junk Film in London's 'Nickelodeons", 1906–1914", *Film History*, vol. 16, no. 2 (2004): 172–197.

80 John A. Hobson, *The Psychology of Jingoism* (London: Grant Richards, 1901), 1–9.

81 See, for example, Steven Attridge, *Nationalism, Imperialism and Identity in Late Victorian Culture: Civil and Military Worlds* (New York: Palgrave Macmillan, 2003), 16–43.

in London consistently displayed, as Jonathan Schneer argues, patriotic, militarist and imperialist sentiments.[82] At that time music halls had bid for a more respectable lower middle-class public and, according to Peter Bailey, became "an agreeable national *alter ego*" promoting a patriotic and nationalist ideology adaptable to all classes.[83] For example, Tommy Atkins, a generic name for the British soldier popularized by Rudyard Kipling, often featured in music hall acts:

> And whether he's on India's coral strand,
> Or pouring out his blood in the Sudan,
> To keep our flag a flying, he's doing and a dying,
> Every inch of him a soldier and a man.[84]

The ideology of music hall recreation migrated into early cinematography. Film pioneers were promoting the imperial theme by creating a visual image of the city which was both modern and also related to tradition and history. Their films were often screened in shows which revealed an imperialist, even military, outlook: for example, during a show including propaganda films against the Boxer Rebellion in China and patriotic films of the Boer War, a showman in London set off blank cartridges in the theatre to increase the realism of the representation – and thus the level of "participation" in imperial military adventures – with haze and the smell of gunpowder.[85] Cities and people in colonized territories were filmed and displayed on film and, conversely, British cityscapes saturated with imperial sights and signs were also filmed and eventually exhibited on the screen in the same cities where they had been filmed. Furthermore, the distribution of these films

82 Jonathan Schneer, *London 1900: the Imperial Metropolis* (New Haven, CT and London: Yale University Press, 1999), 95.

83 Peter Bailey, *Popular Culture and Performance in the Victorian City* (Cambridge and New York: Cambridge University Press, 1998), 150.

84 Quoted in John M. MacKenzie, "Empire and Metropolitan Cultures", in *The Oxford History of the British Empire: The Nineteenth Century*, ed. Andrew Porter (Oxford and New York: Oxford University Press, 1999), 278.

85 "Music and 'Effects' in Cinematography", in *The Showman* (September 6, 1901): 574–75.

in the rest of the world would have served as a powerful and subtle tool for the promotion of Britain's imperial adventure. Early films of London, Manchester or Delhi contribute to the understanding of urban cultures of modernity and imperial power: they offered a visual experience to their original audiences which can be renewed today as we try to understand political and cultural discourses in late Victorian and Edwardian Britain and her empire.

Imperial propaganda and popular culture in early British films

Attempting to make sense of the contemporary reconstruction of capitalist spatiality, Edward Soja has questioned the traditional juxtaposition of history and geography. He has thus suggested a critical understanding of time and space without privileging either, and a vision of the spatiality of social life filled with politics, power and ideology.[86] The present investigation attempts to translate into words the politicized spatiality of the British city in the Age of Empire and aims at an in-depth understanding of the cinematic city as a spatial system by means of an intervention in the discontinuity juxtaposing history and geography.

At the turn of the twentieth century, the meanings of the terms "imperialism" and "empire" were not fixed. During the General Election in October 1900, George Bernard Shaw recognized the centrality of imperialism in British politics and acknowledged its variety of meanings: "The forthcoming General Election will turn, we are told, mainly on the popularity of Imperialism. If this be so, it is important that voters should make

86 See Edward W. Soja, *Postmodern Geographies: the Reassertion of Space in Critical Social Theory* (London and New York: Verso, 1989).

up their minds what imperialism means."[87] The flexibility of the term "imperialism" was also reflected in the historical debate on whether or not empire was a central concern for the British: a number of different views have been expressed by historians, many of whom have emphasized imperialism as the core of British society in the Victorian and Edwardian eras. Historians such as Bernard Porter, for example, have discussed Victorian and Edwardian Britain as a reluctantly imperial society.[88]

During the nineteenth century, ten million square miles and four hundred million people were added to Britain's colonial holdings. David Cannadine has argued that the history of the British Empire and the history of Britain itself should be studied as a seamless whole: the empire was an enlarged and heightened version of British hierarchical and metropolitan society, and both the empire and Britain itself were shaped by conservative attitudes to race, class, and national identity.[89] Cannadine suggests that during the high days of the empire, a homogenizing convergence characterized life in Britain:

> [this convergence was] tied together by a shared sense of Britishness, in which this sense of an ordered imperial society was graded, reinforced, generalized and proclaimed by an elaborate system of honours and titles, and by a pervasive cult of imperial royalty, which surged out from the metropolis to the periphery, and back again.[90]

To date, histories of British cinema have done little to link the early years of film to this shared sense of Britishness which characterized the heyday of the empire. Cultures of imperialism and their relationship with film have been investigated in-depth only in relation to cinema after the First

87 George B. Shaw, *Fabianism and the Empire: A Manifesto by the Fabian Society* (London: Fabian Society, 1900), 1. See also Andrew S. Thompson, "The Language of Imperialism and the Meanings of Empire: Imperial discourse in British politics, 1895–1914", *The Journal of British Studies*, vol. 36, no. 2 (1997): 147.

88 Bernard Porter, *The Absent-Minded Imperialists: Empire, Society, and Culture in Britain* (Oxford: Oxford University Press, 2004).

89 David Cannadine, *Ornamentalism: How the British Saw Their Empire* (London: Penguin Books, 2001), 128.

90 Cannadine, 122.

World War.[91] The era between 1875 and 1914, Hobsbawm suggests, saw the emergence of colonial imperialism. For the first time, the economic and military supremacy of capitalist countries was translated into a formal partition of the world outside North America and Europe into territories administered by western states. Colonialism, Hobsbawm argues, was the most immediately striking aspect of a more general change in world affairs, i.e. the creation of a single international economy; it resulted in the most extraordinary expression of world division into the strong and the weak.[92] Imperialism was of the greatest importance to Britain; the word itself was a novel term describing a new phenomenon, which entered politics in the 1870s and became widely used in the 1890s. Hobsbawm's argument also implies that popular culture contributed to the spread of the term "imperialism" and, more importantly, its message.[93] The cinematograph was a new form of popular leisure which was to become characteristic of the modern urban society of mass culture. Early non-fiction films renewed the imperial project of classification and display of technological novelties, and subjugated people and their places and things, which had earlier been enacted by colonial and industrial exhibitions and photography. The imperial theme thus emerged from films recording life in imperial cities in Britain and cities in colonized territories.

The work of John M. MacKenzie on popular culture and the British Empire reveals the network of connections between the two. Like Cannadine, MacKenzie has argued that imperialism was an essential part of British society, and that in the last three decades of the nineteenth century imperial culture created a world view which was central to the perception the British had of themselves. MacKenzie has also suggested that the generalized imperial vision or world view retained by the British during the Age of Empire embraced cultural hegemony and, according to working-class autobiographies, other memoirs and oral evidence, a common sense

91 See, for example: Martin Stollery, *Alternative Empires: European Modernist Cinemas and Cultures of Imperialism* (Exeter: University of Exeter Press, 2000).

92 Hobsbawm, *The Age of Empire: 1875–1914*, 56–63.

93 Hobsbawm, *The Age of Empire: 1875–1914*, 220, 337–338.

of excessive national pride for all classes. Popular culture, MacKenzie suggests, was an answer to the demand for patriotic themes in a range of media including music halls and cinematography, and a combination of cultural elements which often crossed class lines.[94] Other historians of the Empire have come to similar conclusions. Patrick Brantlinger, who has looked at the relationship between British literature and imperialism, has argued that the imperial discourse informed all aspects of culture and society.[95] John Field has also argued that imperial consciousness was inculcated in forms of travel writing, educational material, journalism, memorials:

> The effort in strengthening and steadying mass character was pursued in a variety of ways, some overt and explicit, most not. The pattern of persuasion basic to general enthusiasm for empire is demonstrated in terms of what rudimentarily educated late Victorians were led to experience vicariously.[96]

Following this train of thought, Robert Storch has suggested that the study of popular culture can contribute to a broader understanding of the history of urbanization by telling us about social, ideological and economic transformations, and can also answer wider questions of social history.[97]

MacKenzie has also investigated the relationship between Empire and metropolitan cultures of different classes and different components of the United Kingdom. MacKenzie has usefully argued that, from the late eighteenth century, cultural practices of entertainment, education, and social activity in Britain reflected and sometimes shaped the instruments of the

94 John M. MacKenzie ed., *Imperialism and Popular Culture* (Manchester and New York: Manchester University Press, 1986), 5–9. On music halls and the imperial theme, see also: Penny Summerfield, "Patriotism and Empire: Music-Hall Entertainment 1870–1914", in *Imperialism and Popular Culture*, 17–48.

95 Patrick Brantlinger, *Rule of Darkness: British Literature and Imperialism, 1830–1914* (Ithaca: Cornell University Press, 1988).

96 John H. Field, *Toward a Programme of Imperial Life: the British Empire at the Turn of the Century* (Westport Connecticut: Greenwood Press, 1982), 83.

97 Robert D. Storch, "Introduction: Persistence and Change in Nineteenth-Century Popular Culture", in *Popular Culture and Custom in Nineteenth-Century England*, ed. Robert D. Storch (London and Canberra: Croom Helm, 1982), 17–18.

political and economic dimensions of imperialism. Cultural expressions of empire were rendered more accessible to a larger public from the second half of the nineteenth century, when entertainment and media became susceptible to new technologies, such as photography and film. These new media began to play a role previously held by engravings and paintings representing the defining moments of the empire and by topical theatre which concentrated on exotic individuals and places of the empire.[98] Shows featuring imperial subjects, as MacKenzie has argued, were hugely popular in late Victorian and Edwardian Britain both in working-class music halls and middle-class theatres, and imperial activity could also be encountered through panoramas, dioramas and cosmoramas exhibited in British urban centres.[99] Early films were exhibited in music halls, one of the most truly distinctive forms of popular entertainment of the nineteenth century.[100] Here, a great variety of acts contained patriotic songs and scenes featuring Britannia, Queen Victoria and the Union Jack. These themes were absorbed by early cinema.

The work of John MacKenzie also challenges the assumption that the British were generally indifferent to imperialism, that it lacked a proper constitutional and cultural expression. This book leans towards the approach of MacKenzie on the basis that Britain's imperial mission and colonial possessions seem to emerge from all aspects of nineteenth- and early twentieth-century British history, culture, society and, as I hope to demonstrate, early films. Whether we browse nineteenth-century issues of *The Illustrated London News*, read newspapers or travel writing from the same age, the imperial theme emerges as the dominant narrative of that time. Both in non-fiction films and in a group of fiction films, imperial narratives played an important role. For example, James Williamson's *Attack on a China Mission* (1900) is a fiction film emphasizing the brutal actions of the Boxer rebels in China and the bravery of the British troops. However, it was mainly in

98 MacKenzie, "Empire and Metropolitan Cultures", 272–275.

99 MacKenzie, "Empire and Metropolitan Cultures", 277–279.

100 See, for example, W. Hamish Fraser, *The Coming of the Mass Market: 1850–1914* (London and Basingstoke: Macmillan Press, 1981).

actuality, topical and interest films that we can trace the connection between empire and urban space. Imperial monuments, exhibitions and parades were filmed in cities and towns, and these images were later exhibited and distributed all over the empire. The influence of imperial propagandist agencies, as MacKenzie suggests, deeply affected the educational system, school textbooks and juvenile literature, youth movements and religious societies, music halls, urban exhibitions and, ultimately, a new-born cinema which embraced imperial nationalism from its very early days, when pioneer filmmakers discovered that imperialism was also profitable.[101] The favourite subjects of topical and actuality films, MacKenzie suggests, reflected the aggressive militarism of juvenile literature and music-hall acts, and the triumphs of the empire over the colonized territories. Royal, military and imperial events, often featuring colonial pageantry, were the subjects of a great number of films presenting a positive image of the empire and avoiding its most controversial aspects.[102]

MacKenzie's argument is characterized by a significant degree of attention for popular culture and, like the present investigation, thus differs from the critiques of imperial culture of Edward Said, which have addressed a somewhat different range of cultural productions. Said's *Orientalism* is important for tackling the complexity of the western representation of the East, combining Michel Foucault's concept of discourse and Antonio Gramsci's idea of cultural hegemony, while transforming Orientalism from an art-historical term meaning a nineteenth-century artistic movement to an ideologically-charged word implying the idea of the "Orient" as a prefabricated myth for western use.[103] However, Said does not sufficiently acknowledge contrasting receptions of Orientalism based on historical and class context and pays insufficient attention to forms of expression outside the context of high culture. Indeed, Said admitted that he had little interest in popular culture, and had concentrated on elite texts:

101 MacKenzie, *Propaganda and Empire*, 34–35.

102 MacKenzie, *Propaganda and Empire*, 68–75.

103 Edward Said, *Orientalism: Western Conceptions of the Orient* (London: Penguin Books, 1995; first published by Routledge and Kegan Paul, 1978); Edward Said, "Orientalism reconsidered", in *Literature, Politics and Theory*, eds Francis Becker et al. (London: Methuen, 1987), 210–229.

> Popular culture means absolutely nothing to me except as it surrounds me. I obviously don't accept all the hideously limited and silly remarks made about it by Adorno, but I must say it doesn't speak to me in quite the same way that it would to you or to my children. I'm very conservative that way.[104]

Michael Hays also investigated the difference between Said's history of imperial culture as it emerges in British high literature and the less uniform and pervasive history that emerges from nineteenth-century popular culture. Both high and popular culture, Hays suggests, can bring a fundamental understanding of the imperialist discourse, and of the ways in which British imperialism acquired the authority and power to impose itself all over the empire. Hays inserts a different construction of the history of imperial culture suggesting that Said's argument that colonial and imperial moments pervaded art in the late nineteenth century should take into account the need to submerge internal cultural differences, and to replace them with a unifying imperial fantasy established in order to stage the empire for all classes.[105] Early cinematography, a product of mass-entertainment exhibited in those music halls dominated by commercial interests and jingoist imperial themes, was a quintessential form of popular culture and an example of the insinuation of imperialism into urban culture. Said's analysis of Orientalism, which has proved an apt tool in the study of other films, is thus a less-adequate instrument in the study of early filmmaking, whose realm is that of popular culture.[106] More specifically, an in-depth understanding of the attempt to domesticate the exotic through a class-based system of honours and rituals must reflect a preoccupation for the ways in which affinities were constructed within the imperialist system and challenge the assumption that the creation of otherness should be seen as the exclusive concern of the British Empire and its cultural agencies.

104 Interview with Said in *Edward Said: a Critical Reader*, ed. Michael Sprinker (Oxford: Oxford University Press, 1992), 246. See also John M. MacKenzie, *Orientalism: History, Theory and the Arts* (Manchester: Manchester University Press, 1995); and Storch, "Introduction: Persistence and Change in Nineteenth-Century Popular Culture."

105 Michael Hays, "Empire: Class, Culture, and the Popular Theatre in the Nineteenth Century", *Theatre Journal*, vol. 47, no. 1 (1995): 65–66.

106 On Orientalism and art cinema, see, for example: Matthew Bernstein and Gaylyn Studlar, *Visions of the East: Orientalism in Film* (New Brunswick, New Jersey: Rutgers University Press, 1997).

Imperial culture was grounded in the urban experience. Nevertheless, the impact of imperialism upon urban space has long been neglected. Felix Driver and David Gilbert, like Jonathan Schneer, have sought to counteract the neglect of the imperial city as an object of study by bringing the experiences of urban modernity and of empire under one umbrella.[107] The chapters contained in Driver and Gilbert's edited book *Imperial Cities* have turned the spotlight on the ways in which empire was expressed in the urban landscape.[108] The present investigation of the ways in which empire was expressed in cinematic cityscapes has also been influenced by Hobsbawm's idea that imperialism was also a cultural phenomenon, and the resulting conclusions are also indebted to the work of James Ryan on Victorian photography, and Thomas Richards and Denis Cosgrove on imperial knowledge and geographical exploration.[109]

107 See Felix Driver and David Gilbert, "Capital and Empire: Geographies of Imperial London", *GeoJournal*, vol. 51, No. 1–2 (2000): 23–32; Felix Driver and David Gilbert, "Heart of empire? Landscape, Space and Performance in Imperial London", *Environment and Planning D: Society and Space*, vol. 16, no. 1 (1998): 11–28; Jonathan Schneer, *London 1900: the Imperial Metropolis* (New Haven and London: Yale University Press, 1999).

108 Felix Driver and David Gilbert eds, *Imperial Cities: Landscape, Display and Identity* (Manchester and New York: Manchester University Press, 2003; first published in 1999).

109 See: Hobsbawm, *The Age of Empire: 1875–1914*; James R. Ryan, *Picturing Empire: Photography and the Visualization of the British Empire* (London: Reaktion Books, 1997); Thomas Richards, *The Imperial Archive: Knowledge and the Fantasy of Empire* (London and New York: Verso, 1993); Denis Cosgrove, *Apollo's Eye: a Cartographic Genealogy of the Earth in the Western Imagination* (Baltimore and London: John Hopkins University Press, 2001).

Empire and the production of cinematic space

Studies of the role of imperialism in shaping the history and geography of the imperial city situate urban landscapes in the context of international cultural and geographical discourses. The urban and the imperial also shaped the production of space in visual culture. Scenes from British and colonial cities had been a popular subject in early panoramas, dioramas, magic lantern shows, photography, and stereoscopic views prior to cinema. Also, Victorians and Edwardians living in big cities and towns could literally see the empire on display at their doorstep when events such as the 1851 Great Exhibition, the 1886 Indian and Colonial Exhibition in London and the 1888 International Exhibitions in Glasgow brought colonial sights to Britain. Dozens of exhibitions took place in a time when both urban and rural spaces of the empire came alive in written and printed media, youth literature, stamps, songs, photography and, ultimately, films. By making the urban experience visible, early films of late Victorian and Edwardian cities contribute to a topographical understanding of modern urban space, and create a visual map of the city and its imperial iconography. According to Erwin Panofsky, iconography can be defined by both the narrow identification of conventional and consciously inscribed symbols and by an intrinsic deep meaning revealing the basic attitudes of the nation, the group or the ideology which produced the icon.[110] Early films of British and colonial cities reveal imperial iconography both in the narrower sense of the word and its deeper stratum of meaning: the imperial iconic landmarks of the city can be seen as consciously inscribed symbols revealing the basic attitudes of the ideology which produced them.

110 Erwin Panofsky, *Studies in Iconography: Humanistic Themes in the Art of the Renaissance* (New York and Oxford: Oxford University Press, 1939), 14–15, and *Meaning in the Visual Arts* (Harmondsworth: Penguin, 1970), 51–81.

As Hobsbawm suggests, British economic progress depended on colonial exploitation.[111] British imperial discourse evoked a world in which less-advanced people were dominated by technologically- and industrially-advanced societies. Stimulated by Enlightenment thought, on the other hand, the imperial discourse entailed a project of world knowledge and exploration which influenced every aspect of social relations, including education, culture, economy and entertainment.

Early films continued the visual display of people and things from distant parts of the globe that characterized nineteenth-century museums and world exhibitions in the United Kingdom. Sites and sights of everyday life and exoticism in the city are here explored in eight chapters. Chapter 1 and 2 trace early cinematic representations of London and focus mainly on the ways in which the imperial theme emerges from images of monumental spaces, urban ceremonies, rites, and exhibitions. In particular, my intention is to map the imperial city through films by focusing on the political core of Westminster and the financial centre of the City. I pay particular attention to the ways in which pioneer cameramen filmed the monumental landmarks of the city and ignored the poorest districts of London and their social problems, thus creating a very partial urban representation of wealth and imperial might. Chapter 3 addresses the ways in which film pioneers portrayed urban space, ceremonies and everyday life in Liverpool and Manchester, with a particular focus on their commercial and industrial dynamism. These cities, where wealth and poverty went hand in hand, emerge from early films, through the selection of themes and places, as unambiguously successful urban spaces. Chapter 4 offers a view on local filmmaking in Scotland, and charts the emergence of imperial sights and signs in Scottish cityscapes in a number of surviving films. In Chapter 5, films of Irish cities are discussed in relation to the hybrid place of Ireland in the British imperial system. Chapters 6, 7 and 8 tackle the ways in which cities in India, East Asia, and Africa were depicted in early films. British cameramen travelled all over the empire filming exotic cityscapes and street

111 See Eric Hobsbawm, *Industry and Empire: From 1750 to Present Day* (London: Penguin, 1968).

life. They depicted colonial urban spaces and exotic cities which played an important role in the Western geographical imagination. The aim of these chapters is to chart a cinematic map of cities in colonized territories in order to draw attention to the ways in which early cinema intervened in the determination of the cultural identities of the colonized and the colonizer, and the boundaries which defined them.

The case studies chosen in this book reflect diverse aspects of urban life at the turn of the twentieth century. London was the heart of the empire and an epitome of urban modernity; it was also the subject of the largest number of non-fiction films depicting one single city. Liverpool, Manchester and Glasgow were defined by a sense of economic dynamism and technological advance; the study of these cities in film contributes to the understanding of the intersection between the experiences of empire and urban modernity. The colonial cities discussed in the third part of this book played a defining role in shaping the European geographical imagination; films of these urban spaces were thus an expression of the growing interest for exotic spectacles in Britain.

This study of the city in the cinema belongs to the wider category of the city and the cinema studies. It presents a cinematic and geographical study of space in films, as opposed to films in space.[112] Nevertheless, I will often refer to the context in which the films were exhibited and address its bearing on its engagement with ideas of empire. According to Lefebvre, space is produced and reproduced through human intentions and shaped by interests of classes and other contending forces. Space is made of geographic form, built environment, rites, everyday life and, Lefebvre suggests, it is produced for domination and to serve a purpose such as to facilitate state power.[113] Cinematic space, as we shall see, reflects and contributes to this process of production.

112 See Mark Shiel, "Cinema and the City in History and Theory", in *Cinema and the City: Film and Urban Societies in a Global Context*, eds Mark Shiel and Tony Fitzmaurice (Oxford and New York: Blackwell Publishing, 2001), 5.

113 Henri Lefebvre, *The Production of Space*, trans. by Donald Nicholson-Smith (Oxford: Blackwell, 1991; first published as *La production de l'espace*, Paris: Anthropos, 1974), 160–164.

PART I

London

CHAPTER 1

Film at the Heart of the Empire

Where early films of London help us to understand how the splendour and the wonders of the modern city are intrinsically related to its miseries and its inequalities.

It is one of the first moving pictures of London: a few horse-drawn carts, men and women walking and bicycling in Hyde Park on a sunny autumn day.[1] The actuality film *Hyde Park Bicycling Scenes* was filmed in November 1896 by Britain's pioneer filmmaker, producer and equipment supplier Robert W. Paul.[2] By watching it and others like it, accompanied by the whir of a Steenbeck table, one may understand what Maxim Gorky might have felt in 1896 when he wrote an article about the first screenings of moving pictures in Russia. The writer from Nizhniy Novgorod wrote about the Lumière cinematograph in the pages of the newspaper *Nizhnegorodskii Listok*, and defined cinema as a place with no sounds or colours: "this is

1 William Friese-Greene had already filmed Hyde Park in 1889. He photographed the park at twelve frames per second on celluloid, but had not succeeded in projecting his film. See Denis Gifford, *The British Film Catalogue: Non-Fiction Film, 1888–1994*, London and Chicago: Fitzroy Dearborn, 2001, 3rd edn), 5.

2 Paul had started business on his own account in 1891, when he began producing electrical instruments and opened a studio in Hatton Garden, in north London. He became involved with cinematography in 1895 when he was asked to make replicas of the Edison Kinetoscope for the Greek entrepreneurs Georgiades and Tragides. During the following years Paul produced hundreds of fictional and factual films, often also directing them, and later earned the title "Father of the British Film Industry." See John Barnes, *The Beginning of Cinema in England, 1894–1901* (Newton Abbot: David and Charles, 1976), 19–36.

not life", Gorky wrote, "but a shadow of life, and this is not movement but the soundless shadow of movement."[3] It is true that at the end of the nineteenth century most films appeared in a monotone grey and, although filmmakers like Georges Méliès started to hand-colour their films as early as 1896, the first successful colour film process was not launched until 1908.[4] Live sound accompanied films since 28 December 1895, when Auguste and Louis Lumière projected motion pictures at Paris's Salon Indien du Grand Café: music, gramophones, lecturers, actors and actresses reading lines behind the screen and, of course, the comments of the audience hardly made of early film exhibition a silent experience. Gorky's view of the cinematograph is thus problematic. The Londoners filmed by Paul, for example, can hardly be associated with what Gorky described as "evil spirits casting the city into eternal sleep."[5] On the contrary, the shadows of movement impressed on celluloid by R.W. Paul and his colleagues offer a significant record of life in the city at the turn of the twentieth century.

Actuality films were a favourite genre in the early days of cinematography. Stephen Kern calls actuality films "one-reel slices of recent history" and emphasizes the ways in which early films of the city were the expression of a historically unique technology for the preservation of the past.[6] For example, in a film like *Hyde Park Bicycling Scenes* we can easily trace signs of the urban culture of its time; the last decade of the nineteenth century was called "The Golden Age of Bicycles" and, in particular, women cyclists in 1896 were still a daring novelty, and mostly seen only in cities or big towns.[7] A popular lyric from Basil Hood's musical *Gentleman Joe* (1895)

3 Maxim Gorky, "The Lumière Cinematograph" in *Nizhnegorodskii Listok*, vol. 4, no. 7 (1896). Translated in *The Film Factory: Russian and Soviet Cinema in Documents, 1896–1939*, eds Ian Christie and Richard Taylor (Cambridge, MA: Harvard University Press, 1988), 25.

4 The earliest successful colour film process was invented in 1906 by George Albert Smith of Brighton, and launched by Charles Urban in 1908.

5 Gorky, 25.

6 Stephen Kern, *The Culture of Time and Space: 1880–1918* (Cambridge, MA and London: Harvard University Press, 2003; first published in 1983), 39, 60.

7 See David Herlihy, *Bicycle: the History* (New Haven: Yale University Press, 2004), 269–271.

describes the enthusiasm of the public for women cyclists: "My eye! Here's a lady bicyclist! Look at her! Look at her! Look at her! Look at her!"[8] A few months before Paul filmed *Hyde Park Bicycling Scenes*, *The Illustrated London News* published an article on the same subject entitled "Cycling in Hyde Park". Old-fashioned *habitués* of Rotten Row (the riding track running along the south side of the Park), said the article, would have found out that "the ladies who used to ride chestnuts, followed by grooms on coal-black steeds, are now pedaling for dear life on bicycles."[9] As this case exemplifies, early actuality films of London were rooted in narratives that occurred off-screen in music-hall acts or illustrated magazines and, as we shall see in this chapter, in a variety of representations of the real offered by nineteenth-century mass media.[10]

In 1913, a journalist writing for *The Times* captured in words the power of this kind of cinematic scene:

> A great part of the enchantment must lie in the fact that the most trivial scene – let us say a meet of coaches in Hyde Park – when cut off from its surroundings becomes for some queer reason significant, even emotional, as it seldom does in reality. Looking up from an arm-chair in a darkened room you see as you have never seen before. The horses and the women and the trees appear on the sheet as if they had nothing to do with the future or with the past, as if the whips would never descend, or the grooms swing up behind, or the horses trot off down the road to Richmond. Let alone the strange way in which isolating something from its context heightens the meaning, there is the sheer excitement and curiosity of the sights themselves.[11]

8 Christopher Pulling, *They were singing and what they sang about* (London: George G. Harrap, 1952), 41.

9 "Cycling in Hyde Park", in *The Illustrated London News*, vol. CVIII, no. 2973 (Saturday, 11 April 1896), 483.

10 My study of this subject is indebted to the work of Vanessa Schwartz, who offered a study of early actuality films made in Paris and their relationship with off-screen narratives like those offered by illustrated newspapers and wax museums. See Vanessa Schwartz, *Spectacular Realities: Early Mass Culture in* Fin-de-Siècle *Paris* (Berkeley, Los Angeles and London: University of California Press, 1998), 190–193.

11 *The Times* (Wednesday, 9 April 1913): 11.

With the emergence of feature-length films, in 1913 short actuality films were no longer as popular as this article might suggest. And yet, the writer well captured the essence of early films of the city, their emotional significance reaching places beyond reality itself. Scenes of everyday life in London were indeed very common in the early days of cinematography and provided an attempt to capture reality while transcending the specific physicality of the places filmed by pioneer filmmakers. The places represented in these films – including exhibitions, museums, zoos and the docks – acquired a larger political and social meaning, often interrelated with ideas of empire. As Felix Driver and David Gilbert suggest, early twentieth-century London was a place in which a great variety of imperial sights could be seen:

> Particular aspects of empire – political authority, commercial power, cosmopolitan consumption, scientific progress, popular display – were represented by different sites such as Westminster, the City, Oxford Street, South Kensington or Wembley.[12]

If London was at the centre of the empire, Driver and Gilbert explain, the empire was at the heart of the urban experience. The modern imperial city, they argue, has been a neglected object of study. Experiences of urban modernity and of empire are usually assumed to belong to two distinct domains: urban sociology and imperial history. Driver and Gilbert investigate the ways in which imperialism shaped modern forms of urbanization, and emphasize the need for a study of imperial London which takes into account both history and sociology.[13] Early films of London as a vehicle of imperial display give us the opportunity to study an image of the city in which the experiences of the empire and of modern urban life constantly intersect. They offer the possibility of studying what Driver and Gilbert define as the public, architectural face of the empire and, in particular, the chance to investigate a cityscape of ceremonies and exhibitions con-

12 Driver and Gilbert, "Imperial Cities: Overlapping Territories, intertwined Histories", 1.

13 Driver and Gilbert, "Imperial Cities: Overlapping Territories, intertwined Histories", 2–3. See also Felix Driver and David Gilbert, "Heart of Empire? Landscape, Space and Performance in Imperial London", *Environment and Planning D: Society and Space*, vol. 16, no. 1 (1998): 11–28.

structed during the heyday of British imperialism.[14] Early films offer a view of a variety of cultural and political expressions of empire, ranging from Queen Victoria's Diamond Jubilee to the display of exotic beasts in Regent's Park and the frenetic activity in the financial heart of the empire, the City of London.

Early actuality films display London as the capital of the largest empire the world had ever seen. This is hardly surprising for, as Edward Said suggests, aesthetics, politics and even epistemology were in the closing years of the nineteenth century inevitably and unavoidably rooted in the culture of the empire.[15] Indeed, the ceremonial spectacle offered by these events displayed London's urban space as utterly linked with the monarchy, performance and with the imperial theme. The proliferation of monumental and commemorative buildings throughout the nineteenth century provided the setting for ceremonial practices which, as David Cannadine argues, were an aspect of international rivalry.[16] Films contributed to make these images of urban splendour familiar to the masses of citizens of the empire: even static cameras, as John MacKenzie has argued, managed to capture the exotic spectacle of imperial troops, horses, and personalities, and anticipated the first successful application of the newsreel technique.[17]

In 1905, Will Baker filmed for the Warwick Trading Company a film entitled *London as the King Sees It*. This film, which is not extant, was filmed from a moving carriage proceeding through the monumental streets of Westminster. The great majority of the cinematic representations of London made before the First World War were filmed "as the King sees" the city or, better still, as the wealthiest parts of society saw the city: the centre of a world-wide empire whose urban pockets of deprivation and neglect were relegated from the screen. Wealthy imperial sites and modern buildings coexisted with the old slums recorded in nineteenth-century photography and literature. Lynda Nead has explained how the dialectic of change

14 Driver and Gilbert, "Imperial Cities: Overlapping Territories, intertwined Histories", 1–3.

15 Edward Said, *Culture and Imperialism* (London: Vintage, 1994), 26.

16 Cannadine, 128.

17 MacKenzie, *Propaganda and Empire*, 68–74.

and continuity characterized the representation of London.[18] Modernity affected London as the result of extremely diverse and unresolved historical processes or, as Nead suggests, as the result of a partial and uneven process of modernization circumscribed by conflicting urban visions.[19] This tension was expressed by the coexistence of the sites of London's poverty and its traditional criminality, its everlasting forms of physical and moral diseases affecting the weakest layers of its population, with the sites of its great buildings and architectural landmarks.

At the turn of the twentieth century, London was indeed a dominant world-city but it was also a socially and geographically divided urban space, with wealthy areas and some of the poorest districts in the country. Early cinema's most conscious act of submission to an imperial vision of urban life lies in its omissions. Charles Booth and Henry Mayhew's earlier shocking accounts of life in the slums can hardly be associated with the wealthy and progressive urban space represented in early non-fiction films.[20] It is true that the last three decades of the nineteenth century saw significant social improvements in the life of London's working classes: by the 1870s sanitary infrastructure like drinking-water, sewers and drains was improving, and morbidity and mortality among London's poor were declining.[21] However, by the end of the century urban pockets of deprivation and neglect, characterized by a catastrophic overcrowding, were still present in the modern capital of the empire. Non-fictional filmmaking in late Victorian and Edwardian Britain did not engage with that spectacle of filth exhibited in prints and photographs representing the poor, the beggars and the most dangerous and insalubrious alleys of the city. In the mid nineteenth century, Gustave Doré and George Cruickshank's grim and atmospheric prints or engravings illustrating novels by Charles Dickens exposed the

18 Nead, "Animating the Everyday: London on Camera c.1900", 68.

19 Nead, *Victorian Babylon*, 4.

20 See Henry Mayhew, *London Labour and the London Poor: the Condition and Earnings of those that will work, cannot work, and will not work* (London: C. Griffin, 1861); Charles Booth, *Life and Labour of People in London* (London and Edinburgh: Williams and Norgate, 1891).

21 Roy Porter, *London: a Social History* (London: Penguin Books, 2000), 332.

darker aspects of low life in London. Later in the century, photography renewed the image of London as a potentially threatening place. Adolphe Smith Headingly and John Thompson's *Street Life in London* (1877), for example, was a collection of social documentary photographs taken in the poorest districts of the city, which focused on the human life of the city and aimed at categorizing the street characters of London.[22] As Caroline Arscott argues, the work of Smith and Thompson seems to indicate a loss of confidence in the modern city.[23] Twenty years later, films suggest an attempt to regain such confidence in urban modernity at the heart of the empire. At the end of a century whose city life was characterized by over-crowded slums, claustrophobic working-class dwellings, and small rooms inhabited by entire families, early cinematographers ignored these spaces and privileged an urban landscape characterized, on the contrary, by open spaces such as parks, squares and orderly wide streets.

In this geographical and historical context, we can see in early cinema a contrast between modernity and historicity: imperial culture aimed at reaffirming or creating tradition on one hand, while promoting technological progress and the emergence of the new on the other. Films both convey a visual record of the fluidity of social interactions in the city and suggest the presence of uneven discourses of modernity and empire. In this chapter I discuss the ways in which pioneer filmmakers depicted London's monumental spaces and its landmarks. The first part of the chapter tackles the representation of such places in everyday circumstances. The second part is dedicated to topical films portraying Westminster and the City during events such as Queen Victoria's Diamond Jubilee.

22 Adolphe Smith Headingly and John Thompson, *Street Life in London* (London: Sampson Low, Marston, Searle, & Rivington, 1877).

23 Caroline Arscott, "The representation of the city in the visual arts", in *The Cambridge Urban History of Britain, vol. 3: 1840–1950*, eds David M. Palliser, Peter Clark, Martin Daunton (Cambridge: Cambridge University Press, 2000), 825.

"Animated pictures, the wonder of all": Ideology and the medium

Victorian London, as William Blake put it, was a "human awful wonder of God", whose sights caused constant marvel.[24] Nineteenth-century literature often addressed the wonders of urban life:

> I long to go through the crowded streets of your mighty London, to be in the midst of the whirl and rush of humanity, to share its life, its change, its death, and all that makes it what it is.[25]

So, in the words of Bram Stoker, Count Dracula tells Jonathan Harker about his intention to move to London. A century after the publication of the novel, in Francis Ford Coppola's *Bram Stoker's Dracula*, a cinematic Dracula meets Mina Harker on the crowded streets of 1897 London and asks for directions to the cinematograph, which he defines as "the wonder of the civilized world." Mina brings him to a music hall where a short comedy is accompanied by a pianist. Dracula proclaims that "there are no limits to science" and looks in bewilderment at the moving images on the screen. Coppola has created a narrative fragment of his own invention, where he pays a tribute to the cinematic essence of the character of Dracula by representing the dawn of cinematography in London. In the late 1890s the cinematograph was a novelty for which the word *wonder* was often used. "Animated Pictures, the Wonder of All" wrote *The Graphic* describing a show at "London's great pleasure resort", Olympia.[26] The form of wonder addressed by Coppola's *Dracula* and by *The Graphic* in relation to the cinematograph is concerned with the technology of the new medium. Technological wonder was not the only form of wonder provoked

24 William Blake, *Jerusalem: the Emanation of the Giant Albion*, in *The Complete Poetry and Prose of William Blake*, ed. David V. Erdman (London: Random House, 1988), 180.

25 Bram Stoker, *Dracula* (London: Penguin, 1993), 31.

26 *The Graphic* (2 May 1895), 26.

by early film exhibitions: the subject matter and the aesthetic of early films were often aimed at provoking wonder in the audience. The study of early cinema is bound to the understanding of the "wonderful" experience of urban space and its modernity, and engages the historical *telos* of the city in the age of empire. Today, in front of urban scenes which took place in London over one hundred years ago, we can experience a sense of wonder similar to that felt by those Victorians who visited the reconstruction of a fifteenth-century street of London at the International Health Exhibition in South Kensington in 1884.[27] The past is subjectively recreated in front of our eyes.

The ideology of a cinematic text, as a form of representation of the real, obscures parts of reality and reveals others. It literally offers an image of society which is created through the spectacle of its achievements and the neglect of its failures. In the case of early actuality films, the dominant imperial ideology caused socially marginal groups to be absent from representation, while the spectacle of its imperial power saturated the screen. The ideological master narrative of London at the turn of the century is defined by its centrality on the map of the empire. This centrality is perfectly expressed by Niels M. Lund in the painting *The Heart of the Empire* (1904), a view of the City of London taken from the roof of the Royal Exchange where we also see Queen Victoria Street bustling with traffic and pedestrians and dominated by the dome of St Paul's Cathedral.[28] In Lund's painting the city extends to the limits of vision and in the early twentieth century the power of imperial London was literally extended to large parts of the globe. Early films renewed this vision of London in a great number of views taken on its streets.

27 See: Annemarie Adams, "The Healthy Victorian City: the Old London Street at the International Health Exhibition of 1884", in *Streets: Critical Perspectives on Public Space*, eds Zeynep Çelik, Diane Favro and Richard Ingersoll (Berkeley, Los Angeles and London: University of California Press, 1994), 203–212.

28 See Stephen Daniels, *Fields of Vision: Landscape Imagery and National Identity in England and the United States* (Princeton, NJ: Princeton University Press, 1993), 13–15.

Shortly before his death on 5 August 1895, Friedrich Engels wrote a new preface for *The Condition of the Working Class in England*, which had first been published in 1844, but translated into English only in 1885 and published in London seven years later. Here he acknowledged that England had outgrown the juvenile state of capitalist exploitation that he described in 1844, but he also implied that the social inequalities which characterized British cities in the 1840s were still acute in the 1890s.[29] In the third chapter of the book, "The Great Towns", Engels described the magnificent sight of sailing vessels on the Thames and the admiration that the view of the river caused in the traveller. However, Engels eventually offered his reader a powerful depiction of a tension between wealth and neglect:

> It is only later that the traveller appreciates the human suffering which has made all this possible. He can only realise the price that has been paid for all this magnificence after he has tramped the streets of London for some days and has tired himself out by jostling his way through the crowds and by dodging the endless stream of coaches and carts which fills the streets. It is only when he has visited the slums of this great city that it dawns upon him that the inhabitants of modern London have had to sacrifice so much that is best in human nature in order to create those wonders of civilisation with which their city teems.[30]

Engels explained how the splendor and the wonders of London are intrinsically related to its miseries, its inequalities and, literally, to its slums and its socially marginal groups. Early filmmakers focused on what Engels bitterly defined as the wonders of civilization. At the turn of the twentieth century these conflicting urban visions still existed, and the dominant ideology was reflected in a cinematic representation of London which was largely concerned with its signs and sights of political and military power.

Jonathan Schneer has discussed the ways in which London's political and municipal authorities carefully stage-managed events such as the return of soldiers from South Africa during the Boer War. The authorities believed that architecture could be made to serve a political purpose and, Schneer continues, displayed the heroes in such a way as to create a mood

29 Friedrich Engels, *The Condition of the Working Class in England* (Oxford: Basil Blackwell, 1971), 360–371.

30 Engels, 30.

of imperialist determination among London's public. As he discusses the case of the City Imperial Volunteers (CIV) marching through London in October 1900, Schneer demonstrates how the itinerary was designed for purposes of spectacle and display.[31] The CIV followed a route from Paddington to the Guildhall passing through the major imperial and architectural landmarks of the capital: Marble Arch, Hyde Park, Piccadilly, St James Street, Pall Mall, Whitehall, the War Office, Marlborough House, Trafalgar Square, the Strand, Fleet Street and St Paul's. Oxford Street and its cheap shops, the insalubrious district of Soho, the slum district of Seven Dials were all carefully avoided. At the turn of the twentieth century the imperial message pervaded everyday life in London, and its built environment contributed to imperial propaganda by staging royal and military parades. As Henri Lefebvre argued, an urban sense of belonging can be found in monumental space and its offering an image of his or her social face to each member of a given society; monuments thus played a key part in supporting the imperial message.[32] Lefebvre writes:

> Thus each monumental space becomes the metaphorical and quasi-metaphysical underpinning of a society, this by virtue of a play of substitutions in which the religious and political realms symbolically (and ceremonially) exchange attributes – the attributes of power; in this way the authority of the sacred and the sacred aspect of authority are transferred back and forth, mutually reinforcing one another in the process.[33]

As much as monuments symbolically underpin urban social membership, films of monumental spaces renew this membership by replacing reality with an ideologically-determined appearance. Monumental space dominates early actuality films of London portraying the imperial and touristic landmarks which were also included in the CIV parade itinerary or described in touristic guides such as the London Baedeker.[34]

31 Schneer, 28–36.

32 Lefebvre, *The Production of Space*, 221.

33 Lefebvre, *The Production of Space*, 225.

34 Karl Baedeker, *London and Its Environs: an Handbook for Travellers* (Leipzig: Karl Baedeker, 1900).

Early filmmakers followed the path charted by London's political and municipal authorities, and offered a fittingly imperial representation of the city characterized by urban landmarks and famous sites. Piccadilly Circus, Trafalgar Square, the Houses of Parliament, Tower Bridge, St Paul's, and Buckingham Palace are constant presences in the work of British filmmakers. The landmarks of the imperial city had already appeared in picture postcards, cigarette cards and juvenile journals which displayed the imperial images and reproduced the landmark buildings of imperial London. Monumental landmarks had often been represented in early photography and related techniques of mechanical reproduction displaying themes which were similar to those of early actuality films. For example, a collection of plastographic views of London from the 1880s includes several places which will later feature in early films: the Victoria Embankment, the Bank of England, Piccadilly Circus, Westminster Abbey, the Strand, Ludgate Circus and St Paul's Cathedral, the Tower of London and Crystal Palace.[35]

The busy streets of London were one of the favourite themes in early films as well as in off-screen narratives such as illustrated newspapers. For example, "Busy London Corners: an everyday Scene at Ludgate Circus" was the title and the subject of a double-spread lithograph published in *The Illustrated London News* in 1896.[36] During the summer of that year many of the places already represented in illustrated newspapers were filmed by Jean Alexandre Louis Promio, the Lumière Company's most prominent cinematographer. Promio shared with the British pioneers a middle-class background, and his films displayed, as Michael Chanan puts it, "the world seen through the eyes of a successful bourgeois."[37] In 1896, Promio filmed five black minstrels dancing and playing musical instruments in Rupert Street in *Nègres dansant dans la rue*. *Danseuses des rues* was filmed in the same year in Drury Lane: in that film three girls dance accompanied by an organ grinder as they are watched by a group of men and boys. More frequently, however, Promio filmed recognizable icons of the city in *Tower*

35 These were stereoscopic pictures of the city; a set of coloured lenses allowed the user to view the images in 3D. The plastographic etchings of London can be found in the James Anderson Collection at the BFI (vol. 13).

36 *The Illustrated London News*, No. 2982, Vol. VIII (Saturday, June 13, 1896): 761.

37 Chanan, 32.

Bridge, *Westminster Bridge*, *Piccadilly Circus*, *Marble Arch* and *Hyde Park*.[38] In 1897, Promio returned from Paris to London and filmed more actuality films in the same fashion: *Départ d'un bateau sur la Tamise*, *Bateau à vapeur sur la Tamise*, *Panorama du Palais de Westminster pris de la Tamise*, *Bateaux sur le lac de St James Park* and *Entrée de Hyde Park*. As we shall see, Promio's selection of subjects charted the way for British filmmakers who later filmed the same places.[39]

Early films of London focused on two distinct focal points: the financial and commercial core of the City of London and the political and cultural centre represented by Westminster. The traffic-choked Strand and Fleet Street, which link together the City and Westminster, were also filmed several times, and the focus was often on the Law Courts. During the closing years of the nineteenth century, pioneer British filmmakers recorded the interaction between modern mass media and ideology by filming London's architectural landmarks. For example, Marble Arch, whose architecture recalls the triumphal Arch of Constantine in Rome and its imperial significance, was filmed by the Chard's Vitagraph Company in 1897. St Martin in the Fields, an urban symbol of the Church of England and therefore the monarchy, was filmed a year later by the Warwick Trading Company.[40] *Pleasure Boats on the Lake of St James' Park* (1898) was the title of an actuality film made by Cecil Hepworth for the Warwick Trading Company and represents the oldest of the Royal Parks of London.

38 Piccadilly Circus was also the subject of Charles Ginner's painting *Piccadilly Circus*, which appears on the cover of this book. For more information on the relationship early cinema and British impressionism see Maurizio Cinquegrani, "Empire and the City: Early Films of London", in Helena Bonett, Ysanne Holt, Jennifer Mundy, eds, *The Camden Town Group in Context*, May 2012, http://www.tate.org.uk/art/research-publications/camden-town-group/maurizio-cinquegrani-empire-and-the-city-early-films-of-london-r1104356, accessed 15 October 2013.

39 The Lumière firm was the first one to exhibit its films all around the word (they went on tour in New York and Bombay as early as in 1896), and their films of London, Liverpool and Dublin were probably the first moving pictures of these cities to be exhibited all over the world. For these reasons, I include them in a book otherwise exclusively concerned with British filmmakers.

40 Titles: *Marble Arch* and *St Martin in the Fields*.

Several actuality films made in late Victorian and Edwardian Britain consisted of a cinematic journey through the streets of London. In 1897, J.W. Rowe's Pictorialograph produced *London*, a series filmed at the Mansion House, Cheapside, Ludgate Circus, Oxford Circus and the Rotten Row in Hyde Park. In 1899 the Warwick Trading Company produced *Hyde Park to the Bank in Three Minutes*, which was composed of ten short scenes including several monumental landmarks of the city, and *London Street Scenes*, a longer series on the same subject.[41] British Pathé, which was established in 1902, produced similar films. One of those, simply titled *London* (1911), features the Bank of England, Tower Bridge, Hyde Park, Piccadilly Circus, Trafalgar Square, Scotland Yard, and the Horse Guards. Charles Urban produced a similar film in 1906: *Hyde Park to Tower Bridge*, which belonged to the series "Streets of London". Urban created arguably the most extensive and detailed travelogues through Edwardian London for the variety of locations: in 1904 his series of scenes called *Living London* included several scenes filmed in iconic locations such as the Embankment, Ludgate Circus, Tower Bridge and the Bank of England.[42] In 1906, he produced *The Streets of London*, a series of scenes including "Hyde Park

41 The series included: "Regent Street from Piccadilly", "Hyde Park Corner Looking West", "The Strand", "Charing Cross Station", "Westminster Clock Tower and St Margaret's", "Piccadilly Looking East", "Piccadilly from Green Park", "Piccadilly Circus", "Waterloo Place", "Trafalgar Square", "Pall Mall Looking East", "Whitehall", "The Strand Looking East", "The Strand Looking West", "The Law Courts – Strand", "Cheapside and Queen Victoria Street", "The Bank of England", "Traffic on London Bridge", "Traffic over Tower Bridge", "Tower Bridge", "Threadneedle and Princes Street", and "Fleet Street Looking East".

42 Ian Christie discovered a version of *Living London* at the National Film and Sound Archive in Canberra in 2006, making this title one of few surviving films of London from that period. The film was shown at the London Film Festival on 24 October 2008. The series included: "Navvies on the Lumber Barges", "Flower Market at Covent Garden", "Covent Garden Market", "Sunday Morning at Petticoat Lane", "The London Coster at Work", "The London Coster at Play", "Costers' Cake Walk at Crystal Palace", "Old Father Thames", "Panorama of the Thames", "Life on the Upper Thames", "Hungerford and Charing Cross Bridges", "Tower Bridge – Raising Same", and "Bank of England and Mansion House."

to Tower Bridge", "Tower of London and Whitechapel", "Five minutes in busiest London" and, remarkably, "From East London Slums to Piccadilly." This film, the only title from that era to contain scenes taken in the slums of London, must have shown a less orderly city and unfortunately is not extant. Four years later Urban produced *Round London with a Camera* and *London and its Life* which included scenes taken in Westminster and the City: "Social and Fashionable London", "Official London", "Political, Legal and Journalistic London", "Financial and Municipal London" and "Historic London and the Ghetto."[43] These films were likely to be a re-edit of *Living London*, perhaps with newly-filmed material, and confirm Urban as a specialist in city material.[44]

As Denis Cosgrove argued, the subject matter of nineteenth-century painted panoramas and dioramas was dominated by images of imperial cities. Actual panoramas of London, Cosgrove explained, were regularly offered to the public at the Regent's Park Colosseum from 1829 and were anticipated by monumental views such as that taken in 1821 by Henry Barker and Thomas Horner from the cupola of St Paul's using Cornelius Varley's graphic telescope.[45] The visual representation of London was then renewed by pioneer filmmakers. Arguably the most complete extant cinematic journey through London at the turn of the century is now known as *Old London Street Scenes*, a film dating from circa 1900.[46] This follows

43 "Historic London and the Ghetto." was possibly filmed in the East End, although London never had a Ghetto in the same sense as in Europe.

44 Other street scenes of London included: *Historic Spots of Old London* (Tyler, 1911), *London of Today* (Gaumont, 1908), *The Streets of London* (Kineto, 1912), *A Day in London* (Gaumont, 1912), *The Traffic in London* (J.B. McDowell, 1913), *Evolution of London Traffic or the Passing of the Horse at the Mansion House* (Hubsch, 1914).

45 Denis Cosgrove, *Apollo's Eye: a Cartographic Genealogy of the Earth in the Western Imagination* (Baltimore and London: The John Hopkins University Press, 2001), 206.

46 There are no available credits for *Old London Street Scenes*, a title of convenience devised by an archivist. I have read most available catalogues of early film companies and, in my opinion, *Old London Street Scenes* is likely to be an uncompleted version of the Warwick Trading Company series *London Street Scenes*, which I have listed earlier in this paragraph. *London Street Scenes* seems to be the only recorded film to

the same route as the CIV marching through London in the October of the same year. Like the authorities that organized the CIV parade, the filmmaker omitted unwelcome sites of urban poverty thus suggesting a conscious and wide-spread attempt to present London as a wealthy and unproblematic urban space. This film represents sites in Westminster and the City, and includes political and juridical landmarks (the Houses of Parliament and the Royal Courts of Justice on the Strand), military icons (the square commemorating the Battle of Trafalgar and the Crimean War Memorial) and places of cultural consumption (the Lyceum Theatre on the Strand). The financial landmarks of the City are given prominence in the second part of the film. The Mappin and Webb building, the Bank of England and the Mansion House appear in *Old London Street Scenes* and were all familiar places in that irregular square mile called the City of London. Eleven scenes filmed with a fixed camera represent as many places; in each of the following paragraphs I explore its locations and draw attention to other films of the same sites.

Hyde Park Corner

The first shot of *Old London Street Scenes* represents horse-drawn vehicles passing by Hyde Park Corner at the west end of Piccadilly, and was filmed just a few yards away from Rotten Row, where Paul had filmed *Hyde Park Bicycling Scene*. The 1900 Baedeker guide to London and its environs informed the reader that "the Park is also a favourite rendezvous of organized crowds, holding demonstrations in favour or disfavour of some political idea or measure."[47] The largest of the Royal Parks was indeed a place devoted to rallies and demonstrations, and by 1850 it was known as the Park of the People. For example, in autumn 1862 several pro-Garibaldi demonstrations took place in the Park, and in 1866 and 1884 members of

include, like *London Street Scenes*, scenes of Waterloo Place. Most of the remaining scenes can also be matched with those of *London Street Scenes*.

47 Baedeker, 333–334.

the Reform League marched to the Park following the defeats of the Reform Bill.[48] Fenian rallies took place each St Patrick's Day throughout the 1860s and 1870s, and in 1880 thirty thousand men and women gathered in the Park to demand state regulation of the laundry industry.[49] Only a glimpse of this social dimension of the Park was offered by early films, when in 1911 Gaumont produced *Suffragettes in Hyde Park*. This film, which is now lost, belonged to a new phase of factual filmmaking: in the early 1910s, British newsreels in the accepted sense were produced by Gaumont and Pathé and presented a broader range of subjects.[50] An earlier film such as *Old London Street Scenes* offers instead an unambiguous view of Hyde Park as an uncontested space; demonstrations were originally relegated off-screen.

Throughout the nineteenth century Hyde Park was the setting for a number of key events in the history of London. In 1851, for example, the Great Exhibition of the Work of Industry of All Nations was staged in Hyde Park. The first Crystal Palace was built in the park to display the latest technological developments in a 990,000 square feet space. This event attracted six million visitors to the capital and was one of the most stunning ventures of the Victorian era.[51] At the end of the century Hyde Park was still associated with ceremonies, monarchy and parades, and became a favourite subject for early actuality films. It had been filmed by Alexandre Promio and several other filmmakers whose moving pictures of Hyde Park have not always survived. Birt Acres filmed *Cycling in Hyde Park* in 1896, and a year later Chard's Vitagraph produced *Hyde Park Corner*. *Rotten Row, Hyde Park* was filmed by Cecil Hepworth in 1898 for the Warwick Trading Company, which also produced *One Sort of Sunday Crowd at Hyde Park*

48 See Jerry White, *London in the Nineteenth Century* (London: Jonathan Cape, 2007), 136, 368–374.

49 Seth Koven, *Slumming: Sexual and Social Politics in Victorian London* (Princeton and Oxford: Princeton University Press, 2004), 165.

50 See Luke McKernan, *Topical Budget: The Great British News Film* (London: BFI Publishing, 1991), 1–5. During the early 1910s at least two more newsreels of political demonstrations displayed a less-imperial urban space: *London: Factory Girls on Strike in Camden Town* (1911) and *London: The Coal Strike March* (1912).

51 See Porter, 253–254.

in 1900, *Coaching in Hyde Park*, which was filmed in 1909 by Will Barker, and, in 1911, *Empire Day*, a film recording the celebration held in the park for this very imperial event. In 1906 Charles Urban filmed *London Fire Brigade at Hyde Park*.

The Park was a complex urban space, and the social and class contrasts visible in the city as a whole can be seen within its 350 acres. Certain areas were inhabited by prostitutes and their clients, and offered an image which was rather different from that of the upper-class Rotten Row.[52] Henry Mayhew gave an account of park prostitution in *London's Underworld*, the fourth volume of his *London Labour and London Poor*, which was published in 1862:

> Park women, properly so called, are those degraded creatures utterly lost to all sense of shame, who wander about the paths most frequented after nightfall in the Parks, and consent to any species of humiliation for the sake of acquiring a few shillings. [...] I was told that an old woman, whose front teeth were absolutely wanting, was known to obtain a precarious livelihood by haunting the by-walks of Hyde Park, near Park Lane.[53]

Just as London's socially marginal groups were absent from representation in early actuality films, the underworld of Hyde Park was never filmed as early filmmakers gave pre-eminence to its most salubrious spaces. Hyde Park Corner was, and still is today, the monumental face of the Park, and *Old London Street Scenes* focuses on its gateway built in 1828 with reliefs inspired by the Elgin Marbles. The neoclassical architecture of the gate, just like the triumphal Marble Arch at the north-east corner of the park, evoked past empires and was an ideal subject for film, a medium born out of imperialism. Other films, such as the Warwick Trading Company's *Lord Roberts' Arrival in London* (1900), contained scenes filmed in the Park and focused on a fittingly imperial event: the appearance of a Boer War hero in Hyde Park.

52 See Judith R. Walkowitz, *Prostitution and Victorian Society: Women, Class and the State* (Cambridge: Cambridge University Press, 1980).

53 Henry Mayhew, *London's Underworld* (London: Spring Books, 1950), 88.

Parliament Square

The second scene of *Old London Street Scenes* is a view across Parliament Square in Westminster filmed from Broad Sanctuary. This shot brings the viewer to the political heart of the city, and features St Margaret's Church and what is arguably the most recognizable monumental building in London: the Big Ben clock tower (over one hundred years and many more establishing shots later, the clock tower is still regarded by Londoners as the most iconic London film location).[54] Several pedestrians and horse-drawn trams can be seen on the screen. The core of Westminster is not as crowded as other districts depicted in this film. As we shall see later in this chapter, it was during processions such as the coronation of Edward VII that the streets of Westminster became the most crowded in London.

The Houses of Parliament, which can be seen in this scene, were in 1900 relatively new: the older Houses of Parliament were destroyed by fire in 1834 and the construction of the new Parliament, designed by Charles Barry, was not finished until 1871. However, Barry's use of Perpendicular Gothic style, which embodied conservative values and evoked English architecture to the period 1350–1550, revealed London's ambivalence towards modernity. By the end of the nineteenth century, it was nevertheless a celebrated architectural landmark and it was filmed several times. In 1896 R.W. Paul filmed *Houses of Parliament*, which exclusively focused on the building. In 1897 Promio filmed the same building from a ferry on the Thames in *Panorama du Palais de Westminster pris de la Tamise*. George A. Smith filmed *Houses of Parliament* in 1897, and Cecil Hepworth made *Panorama of Westminster Palace* in 1898, the same year in which Paul made *Parliament Street*. The nearby Westminster Abbey was filmed in 1911 by J. McDowell for a company called British and Colonial in *A Visit to Westminster Abbey*. These filmmakers portrayed a monumental space filled with symbolic and ritual meaning: by filming the political centre of the empire they contributed to an iconography of political power embodied in London's built environment and its visual representations.

54 "Big Ben most iconic London film location", *Metro* (21 October 2007): 8.

Embankment

Since Alexandre Promio filmed *Dèpart d'un bateau sur le Tamise* and *Bateau à vapeur sur la Tamise*, the river became a favourite subject for early film-makers. The Thames was characterized by a variety of both pleasure and commercial activities. Busy with paddle steamers, ferries, boats and dredgers, it also united the diverse districts of the city. It was the subject of a large number of paintings, photographs and other representations throughout the eighteenth and nineteenth centuries and, as Caroline Arscott argues, it was seen both as a thoroughfare and a landscape for contemplation.[55] In *Old London Scenes* the view of Parliament Square is followed by a shot of the Victoria Embankment taken with a fixed camera from an elevated position, busy with passengers embarking and disembarking from a steamer. A second steamer leaves the embankment in the direction of East London. The 1900 Baedeker guide to London gives plenty of information on the steamboats, including routes, fares and hours.[56] Trips on the river were often both a leisure activity for Londoners and visitors, and also a new way of commuting into the city. *Old London Scenes* thus reveals confidence in urban life at a time when more and more people had time to devote to entertainment and leisure.

In 1898 the same subject was filmed by Cecil Hepworth in *Westminster Steam-Boat Pier* and R.W. Paul's *Panorama of London Streets (Embankment)*. A few years later Charles Urban produced *The Thames: Westminster to Greenwich* (1910), a boat trip from Lambeth Bridge to south-east London. In 1909, two boat trips were filmed by Gaston Quiribet for Hepworth in *The Thames in Winter* and *Moonlight on the Thames*. In the same year, Will Barker filmed for the Warwick Trading Company a more fittingly imperial image of the Thames: *The Great Thames Pageant* represented the Home Fleet arriving in the Port of London. Urban filmed the same event, and his film included scenes of battleships and submarines. In early actuality films the river was filmed giving particular attention to the municipally-owned

55 Arscott, 817–820.
56 Baedeker, 63–64.

steamers, which were inaugurated in 1905, its modern bridges and their lively traffic scenes. In 1896 Westminster Bridge, built thirty-four years earlier, was filmed by Promio, R.W. Paul and Esmé Collings. In the same year, Paul also shot a busy scene on Blackfriars Bridge, *Rush Hour Record on London Bridge* and *Traffic on Tower Bridge*. The latter represented the most recent and original bridge on the Thames, completed in 1894 and destined to become another iconic London location. The City of London Corporation, the municipal governing body of the City of London, commissioned the building of Tower Bridge when the East End became so densely populated that a new bridge east of London Bridge became necessary. Early actuality films representing the bridge and the steamers thus exemplify municipal power in late Victorian London. Tower Bridge, which represented an attraction for Londoners, tourists, filmmakers and audiences alike, was also filmed in 1896 by Promio and Birt Acres in *Pont de la Tour* and *Tower Bridge Working*. A few years later, Will Baker filmed *Panorama of the Tower of London and Tower Bridge* for the Warwick Trading Company (1904).

Waterloo Place and Trafalgar Square

Two of the following three scenes of *Old London Street Scenes* bear a straightforward military and imperial message, as they represent monumental sites celebrating the military achievements of the empire. The first of the two shots briefly depicts John Henry Foley's Crimean War Memorial in Waterloo Place, located at the junction of Lower Regent Street and Pall Mall.[57] The image of a monument worshipping the heroes of an earlier war put forward a military message during the Second Boer War in South Africa (1899–1902). *Old London Street Scenes* continues with a brief

57 The memorial was considerably changed by Arthur George Walker in the 1910 and unveiled in 1915. The original statue of Sidney Herbert, who was responsible for the War Office during the Crimean War, became part of a bigger memorial, including the statue of Florence Nightingale.

shot of Charing Cross station. This is followed by a view of Pall Mall and Trafalgar Square. Horse-drawn carriages move towards the foreground, and the National Gallery can be seen in the background.

Trafalgar Square and Nelson's Column are among the most outstanding monumental landmarks in London and the most popular destinations for its visitors. Thanks to its location at the intersection of Pall Mall to the west, Whitehall to the south, and the Strand and Fleet Street to the east, Trafalgar Square was the meeting point of the British monarchy, government, and commerce. The monument was completed in 1845 and in the nineteenth century, just like today, it was considered the heart of London, the midpoint between the City and Westminster. Rodney Mace explained that the square was a contradictory space, designed as an imperial space and later established as a site of political demonstrations, whose imperial iconography provided those who challenged the established order with a target.[58] The first moving picture of the place was made in 1890 by Wordsworth Donisthorpe and William Carr Crofts, two British inventors and pioneers of photography.[59] Ten frames from their film representing water surging from one of the fountains framed against a background of the National Gallery, with pedestrians and horse-drawn omnibuses, have survived. In 1897, R.W. Paul filmed the same subject in *Fountains*. Throughout the 1890s London County Council pursued the control of water, and in 1903 the Metropolitan Water Board was founded, thereby bringing under a single public body nine private water companies supplying water to London.[60] In these films, Trafalgar Square fountains could also

58 Rodney Mace, *Trafalgar Square: Emblem of Empire* (London: Lawrence & Wishart, 2005), 155–203.

59 Their film was photographed at 10 frames per second on celluloid, with a camera called Kinesigraph. See Gifford, *The British Film Catalogue: Non-Fiction Film, 1888–1994*, 3.

60 The New River Company, the East London Waterworks Company, the Southwark and Vauxhall Waterworks Company, the Company of Proprietors of the West Middlesex Waterworks, the Company and Proprietors of Lambeth Waterworks, the Governor and Company of Chelsea Waterworks, the Grand Junction Waterworks Company, and the Staines Reservoirs Joint Committee. See: Andrew Saint (ed.), *Politic and*

be seen as a demonstration of municipal-owned water supplies, with their constant jets of water aimed at communicating opulence and the power of the municipal authority.

On 21 October 1897 the celebrations for the ninety-second anniversary of the Battle of Trafalgar were filmed by William Dickson for Mutoscope and Biograph (*The Nelson Column*), Cecil Hepworth for the Warwick Trading Company (*Trafalgar Day in London*), George Albert Smith (*Nelson's Monument*) and W. Watson & Sons (*Trafalgar Day*). In 1909 Gaston Quiribet filmed *City of Westminster* for the Hepworth firm. This film is a phantom ride filmed from a car on the streets of Westminster, from Whitehall to St Martin's place. Trafalgar Square not only appears as the physical centre of the city, but also as one of its most lively places. *City of Westminster* emphasizes its importance in the urban environment by documenting the presence of a great number of automobiles, omnibuses, trams and pedestrians dwelling in the area. Nelson's Column is filmed here with a vertical shot from its base to the statue of Admiral Nelson facing towards Westminster and along Pall Mall.

The square was also, and like Hyde Park still is, a venue for political demonstrations: during the 1880s the Labour movement, in particular the Social Democratic Federation, held rallies in Trafalgar Square. On 8 February 1886, a major rally against unemployment led to riots in Pall Mall. Later that year, on 13 November, thousands of people marched to the square to demonstrate against coercion in Ireland; due to the violent reaction of the police, that day would become known as "Bloody Sunday".[61] At the beginning of the twentieth century, suffrage reform became a popular issue addressed by the square's demonstrators. One of the demonstrations organized by the National Union of Women's Suffrage Societies was the subject of a film produced by the Warwick Trading Company on 13 June 1908: *The Suffragette Demonstration* represents the Women's Suffrage March in one of London's major venues in for political demonstrations. This

---

People of London: the London County Council 1889–1965 (London and Roncervete: Hambledon Press, 1989).

61 White, 368–379.

film is something of an exception: demonstrations and riots were largely absent from the cinematic representations of the square commemorating Nelson's victory at the Battle of Trafalgar in 1805. Trafalgar Square was an ambivalent space embodying elements of both power and disorder but the cinematic image emerging from most films here listed reflected its iconic identity as symbol of political power and civic pride.

The Strand

The two following shots of *Old London Street Scenes* represent Fleet Street and the Strand: in the first we can see the Law Courts, while the second shot was taken on a crowded traffic-choked part of the Strand near the Lyceum Theatre, with St Mary le Strand in the background. Its crowd and its traffic suggest the importance of the Strand in London's topography at the turn of the century. Patrick Joyce has investigated practices of movement in the city and explained the role played by the Strand:

> The living stream of London is above all that which flows in and out of the city every day, increasingly the commercial City itself, and which is in fact to be encountered most clearly at the place where this tide ebbs and flows most, fittingly enough called the Strand.[62]

Joyce has suggested that the nineteenth-century city was experienced in new kinds of ways and argued that the stream of the city, including the ritual of going to work and of leisure, was part of the urban grip on the city's inhabitants.[63] This scene from *Old London Street Scenes* records the ritual of going to work through a vital artery of London, where thousands of people walked to the City from Charing Cross station. Physical business transactions between the City and Westminster depended on communication between these two areas. The importance of the Strand was reflected

62 Patrick Joyce. *The Rule of Freedom: Liberalism and the Modern City* (London and New York: Verso, 2003), 227.

63 Joyce, 203–227.

in the city authorities' plans to redevelop the thoroughfare by demolishing the narrow alleys near it. In 1900, a portion of the slum property between the Strand and Holborn (such as the buildings occupying Holywell Street) was demolished to make way for Aldwych and Kingsway.[64] In contrast to narrow alleys like Holywell Street, wide streets such as Kingsway and the Strand allowed public free passage on a greater scale. An aspect of international rivalry and inspired by Georges-Eugène Haussmann's rebuilding of Paris, the planning of these streets can be seen as London's answer to the wide boulevards of other imperial capitals, such as Paris and Vienna. As *Old London Street Scenes* exemplifies, filmmakers seeking an imperial image of the city were thus attracted to wide thoroughfares like the Strand. On the other hand, the slum district running north of and parallel to the Strand was never filmed by early filmmakers who consistently avoided the insalubrious districts of the city.

The City

The last three shots of *Old London Street Scenes* bring the viewer into the heart of the Square Mile. A view of the Mappin and Webb building at Poultry is filmed from an elevated position looking towards Cannon Street; the bell tower of St Mary le Bow can be seen in the background. This shot is followed by two street scenes outside the Bank of England and the Mansion House. Union flags are waving on horse-drawn omnibuses, and the dense traffic in Queen Victoria Street and Threadneedle Street reflects the identity of the City as the commercial and financial nexus of the empire, its frenetic source of economic power. The activities undertaken in this space contributed to the determination of imperial London, as Schneer suggests, by cementing the City's role as a financial and commercial metropolis and encouraging the government to pursue an imperial policy in order to reap the benefits of imperialism.[65] The film ends with a

64 See Nead, *Victorian Babylon*, 163.
65 Schneer, 64–92.

shot of the area in front of the Lord Mayor's residence, the Mansion House, taken from an elevated position (probably from a window of the Mappin and Webb building). If each of the eleven scenes composing *Old London Street Scenes* reveal how crowded and frenetic was late Victorian London, these three shots of the City reveal the busiest district of London, with the highest number of people and vehicles of various kind on the streets. In particular, the final shot, by far the longest shot in the film, displays an endless flow of carriages, people and horse-drawn trams with banners and poster advertising Nestlé and Mellin's products.

The significance of the City is also revealed by the built environment. As Richard Dennis suggests, the imperial character of the City was conveyed equally by building styles evoking past empires, and by celebrations such as Queen Victoria's Diamond Jubilee.[66] Like Marble Arch in Westminster and the Mansion House with its Palladian style, the classical architecture of the Bank of England evokes the Roman Empire.[67] Films of these buildings thus suggest the theme of empire in the cinematic image of London. The great buildings of the City and scenes such as that painted by Lund in *The Heart of the Empire* appeared in several other films. In 1898, William Dickson filmed *London's Traffic at the Mansion House* for Mutoscope and Biograph. The Square Mile was also filmed in 1909 by Gaston Quiribet for the Hepworth firm: *City of London* was to be paired with the extant *City of Westminster*. In 1903 the Walturdaw Company produced a film called *Busy London: Traffic Passing in Front of the Bank of England and Mansion House*. This film was shot during peak hours from the Mappin and Webb building at Poultry and offers a view of the frantic activity of the Square Mile.

Thus, *Old London Street Scenes* displays the most well-known architectural landmarks of London. This film reveals imperial iconography both in the narrower sense of the word and that wider stratum of meaning

66 Richard Dennis, "Modern London", in *The Cambridge Urban History of Britain, Vol. 3: 1840–1950*, eds David M. Palliser, Peter Clark, Martin Daunton (Cambridge: Cambridge University Press, 2000), 130.

67 Simon Bradley and Nikolaus Pevsner, *London 1: The City of London* (London: Penguin), 317–321.

which, according to Erwin Panofsky, can be defined by an intrinsic deep meaning revealing the basic attitudes of the nation, the group or the ideology which produced the icon.[68] Several places represented in *Old London Street Scenes* are inscribed symbols of empire revealing an expression of the basic attitudes of the society and the ideology which produced them. The urban landmarks which appear in this and similar films constituted the official face of the urban centre of a world-wide empire. The city which emerges from these films does not only display a variety of imperial symbols, such as the Crimean War Memorial and Nelson's Column, but it is also presented as a modern urban space characterized by free movement, activity and crowds. *Old London Street Scenes* informs us of the ways in which early films renewed the visual representation of monumental spaces of London inscribed into the everyday urban experience of the empire. In the following section, I discuss the cinematic representation of Westminster and the City during imperial events and focus on the ways in which the extraordinary replaced the everyday in films of processions such as that for the sixtieth anniversary of Queen Victoria's accession to the throne.

Crowds, processions and imperial events in early topical films

"What is the city but the people?" asks the tribune Sicinius in William Shakespeare's *Coriolanus*; "True, the people are the city" reply the citizens of Rome. And it is also true that films offering an image of the city often included large crowds. The depiction of London which emerges from the street scenes is one of a chaotic aggregation of activities, forms of urban transport, posters and banners advertising products and services (in particular advertising for Nestlé products appears on many omnibuses). Most remarkably, it is a depiction of people and mobility. From the mid nineteenth century, artists like William P. Frith and George E. Hicks took the

68 Panofsky, 51–81.

urban crowd as their subject in paintings and inaugurated a long-lasting fascination with the crowd in visual representations of London.[69] In *Old London Street Scenes* the Strand, Queen Victoria Street and Threadneedle Street in the City are choked with traffic and frenetic crowds of individuals pursuing their work and other daily activities. Crowds were represented in different ways in early actuality films of London, and indeed the word "crowd" appeared in several titles. R.W. Paul's *A Crowd in the City* and Esmé Collings' *Crowded Streets in London* were both made in 1896 and are now lost, but these titles suggest an attention to what Edgar Allan Poe called "the very denseness of the company" in the principal thoroughfares of the city.[70] In this section, images of order and disorder are used to investigate two different London crowds: Londoners working and shopping at Petticoat Lane Market on one hand and the public of imperial events on the other. In particular, we are going to look at those films of Queen Victoria's Diamond Jubilee which depict the city at its imperial apogee. The Jubilee films, as well as other films of urban and imperial events, also give us the opportunity to discuss the cinematic representation of ceremonial practices in London at the turn of the century. Before this, and in order to emphasize the diversity of the crowds of London, I shall address *Petticoat Lane*, a remarkable film depicting a popular district in East London, and an example of a cinematic counter-tendency focusing more on popular districts of the city and their communities, and less on monumental spaces and imperial sights.

Petticoat Lane, which was filmed in 1903 by an unidentified filmmaker and is one of the most interesting actuality films representing street life in early Edwardian London. Most actuality films of London from the late Victorian and Edwardian eras are not concerned with the city's poorest districts. This is what makes *Petticoat Lane* a very special film (as we shall see, other films of markets were made but have not survived to today). From a formal point of view, in *Petticoat Lane* the filmmaker employed a number

69 See Arscott, 820–827.

70 Edgar Allan Poe, "The Man of the Crowd", in *The Fall of the House of Usher and Other Writings* (London: Penguin), 132.

of panning and panoramic shots representing the crowd at Petticoat Lane Market in Whitechapel, East London. *Petticoat Lane* was probably filmed on a Sunday, when the market extended over the surrounding streets, and was busier than on any weekday. The market took place, and still does, on Wentworth Street and Middlesex Street, which was widened in 1900 following demolitions which gave the market more space. In *Petticoat Lane*, the market is crowded with men and boys (very few women are visible on screen) and sellers standing above the crowd, waving their products and promoting the goods they are selling, men's clothes and watches in particular. As David Feldman has argued in discussing the presence of Jews in London at the turn of the twentieth century, certain streets of the East End, such as Wentworth Street and Middlesex Street, seemed like "a piece of eastern Europe thrown down in the capital of the world's largest empire."[71] In 1898, the market was described as follows in the guide *Through London by Omnibus*:

> A walk through this curious bazaar on a Sunday would give one sensations enough to last for some time. The motley crew speaking and shouting almost every human language, the diversity of the costume, the curious jumble of the goods for sale on the stalls, and smell of fried fish, garlic, and the like, make up an experience that requires to be gone through to be appreciated.[72]

Petticoat Lane is a portrait of this exotic urban space, and documents the presence of Russian and German Jews in the East End, as several store signs outside Middlesex Street bear Jewish names: at number 52 there is an "R. Isaacs Anteka" (*anteka* is Russian for pharmacy), "S. Ginsberg" names an unidentified business, and the sign "This is Levy's" advertises the famous Levy's rye bread. Other signs advertise "Idris Mineral Waters" and "Soda

71 David Feldman, "Jews in London: 1880–1914", in *Patriotism: the Making and Unmaking of British National Identity: Minorities and Outsiders*, eds Raphael Samuel (London and New York: Routledge, 1989), 212.

72 C. Hodder, *Through London by Omnibus; A Chatty Guide to the Principal Sights of London* (London: Hodder, 1898), 63. See David Gilbert, "London in all its glory – or how to enjoy London: guidebook representations of Imperial London", *Journal of Historical Geography*, vol. 25, no. 3 (1999): 286.

Milk." The streets where Petticoat Lane Market took place were also frequented and inhabited by other ethnic groups. In *Petticoat Lane*, the Irish presence in this multi-cultural district is confirmed by the sight of a large travel agency advertising fortnightly departures to Belfast, Dublin, Cork, Kilkenny and Waterford. The presence of a large Jewish community in the East End had created greater social tensions in the area, as they were perceived as further competition for jobs and housing, and there were often riots against them.[73] Notwithstanding this, in *Petticoat Lane* the East End crowd is cheerful. Just as pockets of inner-city poverty were largely absent from representation in early actuality films, the working-class district in East London was filmed without any hint to the social tensions affecting this area.

This actuality film offers a vivid representation of Petticoat Lane Market's communities. Other films of London's markets from the turn of the twentieth century are now lost. R.W. Paul also made a film titled *Petticoat Lane* in 1896. Three years later William Dickson filmed *The Lane on Sunday Morning* for the Biograph Syndicate. In 1904, Charles Urban's series *Living London* included "Flower Market at Covent Garden", "Covent Garden Market", and "Sunday Morning at Petticoat Lane." In the same year, Paul filmed London's oldest street sellers in *Street Hawkers*. One year later Cecil Hepworth made *Sunday in Petticoat Lane*, and in 1906 British Gaumont produced *Spitalfields Market* and *Sunday Morning in Petticoat Lane*. In 1910, Charles Urban released a film called *London Markets* which included "Smithfield", Billingsgate" and "Covent Garden." The latter, the most celebrated of London's fruit and vegetable markets, was filmed again in 1912 by J.B. McDowell for the B&C: *Covent Garden*. *Petticoat Lane* offers an image of Londoners dealing with everyday activities and spaces. More broadly, films like *Petticoat Lane* offer a representation of a chaotic crowd, a disorganized agglomeration of people which strikingly differs from the masses represented in topical films of rituals such as imperial parades, marches, coronations and funerals. Those films represent sights of imperial power and offer images of regular movements across the city, organized routes, and orderly crowds lined by soldiers.

73 See Karl Beckson, *London in the 1890s: a Cultural History* (New York and London: W. Norton & Company, 1992), 7.

Films of parades and processions give us the opportunity to understand the imperial city in terms of spectacle and urban spectatorship. Other media, such as illustrated magazines, had already exploited official events taking place in London. Browsing the 1896 issues of *The Illustrated London News*, for example, one finds a great number of articles devoted to parades, visits and funerals of members of the Royal Family, nobles, military heroes or politicians. Articles, illustrations and photographs of the funeral of Sir Frederic Leighton, president of the Royal Academy, in St Paul's Cathedral occupied the cover and inside pages of *The Illustrated London News* in the issue released on 8 February of that year.[74] Urban events and their representations in films and other media often magnified the imperial ruling class and gave audiences an opulent spectacle on film. The ceremonial practices and orderly parades which migrated as a subject into early films had characterized the use of London's public spaces for centuries. Steven Mullaney, who discussed the power of ritual in sixteenth-century London, argued that the particular theatrical valence of rituals dramatized the city's image of itself and made it inseparable from the physical body of the community. In other words, according to Mullaney, Elizabethan London was shaped by a number of rites through which a ceremonial social order defined itself.[75] Both in sixteenth- and nineteenth-century London, as Ian Munro suggests, innovative ceremonial practices coexisted with older rites, and they contributed to produce urban legibility. At the end of the nineteenth century, Munro continues, urban rituals such as parades were staged to promote the imperial theme and transformed everyday space into an urban space of transparent meaning.[76] Imperial parades mostly took place in central London, from Westminster to the City, through the Strand. These events contributed to the development of the idea of London as "home", in the sense in which, as Raymond Williams suggests, "home is a memory and an ideal" offered by the images of "the powerful, the prestigious and the con-

74 *The Illustrated London News*, No. 2964, Vol. CVIII (February 8, 1896), 1, 164.

75 Steven Mullaney, *The Place of the Stage: License, Play and Power in Renaissance England* (Chicago: University of Chicago Press, 1988), 10–15.

76 Ian Munro, *The Figure of the Crowd in Early Modern London: the City and its Double* (New York: Palgrave Macmillan, 2005), 24–26.

suming capital."[77] In his discussion of urban and rural landscapes, Williams defines nineteenth-century London as an imperialist urban space whose grandeur coexisted with the creation of centres of poverty and misery in the East End.[78] The latter were untouched by parades and processions, just as they were neglected by filmmakers.

As I have argued in the previous section of this chapter, early actuality films representing London's monumental spaces were filmed on an urban route which follows the itinerary of parades such as the return of the CIV and, by doing so, they privileged the monumental spaces of the city. Many other ceremonies and processions took place on the streets embracing the centre of London, from Hyde Park to the Guildhall. Imperial and political rites such as the coronation of Edward VII and George V were filmed by early filmmakers and emerged as some of the most remarkable products of the early British film industry. In these films, the streets of Westminster and the City, with their ceremonial function, were visually very different from Petticoat Lane and its multicultural crowds of Londoners. In the ceremonial spaces of London, distinct cultural groups were displayed during processions as sights of imperial domination in the form of colonial troops.

The first and most striking event of imperial significance to be filmed by several cameramen, representing both British and foreign companies, was Queen Victoria's Diamond Jubilee in 1897. Between 19 and 24 June, London celebrated the sixtieth anniversary of Queen Victoria's accession to the throne with a number of ceremonies, speeches and official processions. The Diamond Jubilee was staged as an attempt to make all British citizens feel proud of their country and empire. As Denis Judd suggests, its imagery and symbolism were exploited by imperial propagandists to "invent tradition" in order to highlight British imperial achievements and to bolster national morale.[79] Driver and Gilbert have discussed the ways in which the image of the empire was exploited on the Jubilee parades and other royal processions in other to make London even more fittingly imperial:

77 Raymond Williams, *The Country and the City* (London: Chatto & Windus, 1973), 281–282.

78 Williams, 283.

79 Denis Judd, *The British Imperial Experience from 1765 to the Present* (London: Phoenix, 2001), 130–153.

> With the reinvention of the monarchy in the late 19th century, state ceremonial processions through the City became more common and more spectacular, culminating in the Diamond Jubilee procession of 1897, when Bank Junction was transformed by giant allegorical paintings, drapes, and imported evergreen trees.[80]

On June 26, *The Illustrated London News* published three illustrations representing the illuminations throughout the city, and emphasized how both technological progress and imperial prestige were celebrated by thousands of gas lamps outlining buildings such as the Mansion House.[81] The newborn cinematograph contributed to the celebration of the imperial message during the celebrations for Queen Victoria. They depicted the Jubilee by focusing on celebratory images such as those reproduced on the pages of *The Illustrated London News* and brought images of the event to those citizens of the empire who had not gathered in London for the Jubilee. The climax of the Jubilee was the royal procession to and from St Paul's Cathedral on 22 June, when an estimated crowd of three million people gathered in the streets of the central districts of the city. London's crowds viewing parades and celebrations such as the Jubilee were conscious of living at the heart of the empire and, with the media, local and national authorities contributed to forming imperialist-minded crowds.

More than any other event, the Diamond Jubilee contributed to fixing what Thomas Richards defines as "a ready-made image of royal power."[82] In his discussion on the semiotic consolidation of capitalism in nineteenth- and twentieth century Britain, Richards has explained how the Jubilee was staged as a celebration of Britain's imperial achievements aimed at displaying domesticated sights of otherness:

> The organizers rolled out an imitation oriental rug in the form of a carefully contrived allegory of Empire that transformed its distant and often menacing otherness into the familiar sight of obedient ranks of soldiers under direct British military command. Riding high in the wake of hussars from Canada, carabiniers from Natal,

80 Felix Driver and David Gilbert, "Heart of Empire? Landscape, Space and Performance in Imperial London", 18–19.
81 *The Illustrated London News*, No. 3036, Vol. CX (Saturday, June 26, 1897): 888.
82 Thomas Richards, *The Commodity Culture of Victorian England: Advertising and Spectacle, 1851–1914* (Stanford: Stanford University Press, 1990), 71.

> camel troops from north Borneo, white-gaitered Jamaicans, Cypriots with fezzes, and Guiana police in peaked caps, the domesticated queen at last became the charismatic center of all the phantasmagoria of Empire, a symbolic exemplar of what the *Daily Mail*, in headlines of gold ink on June 21, 1897, called "GREATNESS OF THE BRITISH RACE."[83]

On June 12, *The Illustrated London News* published a double-spread illustrated map of the route to be taken by the procession. Queen Victoria would leave Buckingham Palace at 11.15 a.m. and pass through Constitution Hill, Piccadilly, St James's Street, Pall Mall, Trafalgar Square, the Strand, Fleet Street, and Ludgate Hill. She would then arrive at St Paul's Cathedral, where a short ceremony would take place. The Queen's progress would continue via the Mansion House and the Bank of England across London Bridge and through South London, before returning over Westminster Bridge, past the Houses of Parliament, to Buckingham Palace.[84] The procession on June 22 passed by all those monumental landmarks which often featured in early actuality films portraying everyday life in the imperial city. Months prior to the event *The Times* described the plan to turn London into a mirror of the empire during the celebration:

> The Indian Empire will be represented in symbolical fashion in Piccadilly at the top of St James's-street, the Services would be similarly represented in Pall-mall, Canada in the Strand, the whole British Empire at Ludgate-hill viaduct, Australasia at the west end of Cheapside, Africa in King William-street, and the British empire a second time in the High-street, Borough.[85]

The procession saw almost every British cinematograph firm and several cameramen from abroad represented along the route.[86]Arrangements for the filming the Jubilee were described on the pages of *The Photogram*:

83 Richards, 116.

84 *The Illustrated London News*, No. 3034, Vol. CX (Saturday, June 12, 1897).

85 *The Times* (Wednesday, 31 March 1897): 4.

86 These included: Cecil Hepworth for the Warwick Trading Co., the Lumière firm, William Dickson for the Mutoscope and Biograph, Chard's Vitagraph, Prestwich Manufacturing Company, Birt Acres, R.J. Appleton, London Stereoscopic Co., G.A. Smith, Adolphe Langfier for Velograph Syndicate, Haydon & Hurry,

> At the time of writing every kinetographer is busy with arrangements for making views of the Diamond Jubilee procession and some of them have already (a fortnight in advance) booked large orders for the resulting positives. Some of the firms make quite a mystery of their intentions, but we know what some of them are doing. The most extensive arrangements of which we have heard are those of J.W. Rowe & Co., who will have six operators at work at different points, including one facing the service on the steps of St Paul's. J. Neville Maskelyne will photograph the service from his own grand-stand. The Velograph Syndicate has made arrangements for two or three operators at good points; Dr J.H. Smith &Co (of Zurich) will have two fine positions in the yard of Charing Cross-station, facing the procession as it comes from Duncannon-street into the Strand; and the Moto-Photo Supply Co. will have one of the best position on the whole route for commanding the procession itself, at the corner of Parliament Street and Great George Street, overlooking the northern slope from Westminster Bridge. The Lumières (Fuerst Bros.) will have more than one experienced operator on the route, and we understand that both Birt Acres and W. & D. Dowey have received valuable official concessions in the matter of positions.[87]

Filmmakers captured an imperial ritual in films which became factors defining the production of urban space. This social space, Henri Lefebvre argued, is produced and reproduced in connection with forces of production defined by the mediation of factors "within knowledge, within ideology or within the domain of representation."[88] In order words, the forces of production take over a space which is not empty, neutral, nor solely determined by nature. In films of the Jubilee, knowledge emerges from a practice whose symbolic elements have their source in history. The Jubilee can be contextualized in what Pamela K. Gilbert calls "an anxious sense of London's historical importance and relation to both modernity and history" which characterized the nineteenth-century city.[89] The answer to such anxieties was revealed by the intersection between politics and architecture

---

W.C. Hughes, Ludgate Kinematograph Co., Alfred Wrech and Sons, John Le Couteur for the Photographic Association, Watson and Sons, Motorgraph, British Cinématograph Co., Moto-Photo Supply Co., J. Neville Maskelyne and J.W. Rowe's Pictorialograph.

87 *The Photogram*, vol. 4, no. 43 (1897): 214.

88 Lefebvre, *The Production of Space*, 77.

89 Pamela K. Gilbert, "Introduction: Imagining Londons", in *Imagined Londons*, eds Pamela K. Gilbert (New York: State University of New York Press, 2002), 3.

which led to the multiplication of imperial symbols, statues, monuments, and urban redevelopment projects such as the creation of Aldwych and Kingsway. The answer to that anxious sense of London's historical importance was also given by celebrations like that Jubilee and the temporary transformation of the urban landscape which accompanied them.

The places included in the Jubilee parades bear witness to the existence of a unitary code of the city which combines the city's reality with its symbolic, practical and imaginary ideal. The theatricality of the city's monuments, particularly as their imperial significance is celebrated by parades and decorations, is coextensive with its everyday life and functions (such as the activities represented in *Petticoat Lane*). However, the imperial sight offered by the Jubilee overshadows the everyday signs and sights of the empire displayed in actuality films representing quotidian activities. The Jubilee films were screened within a few days in the capital's music halls, where patriotic plays were also staged, and were eventually distributed and exhibited in all the corners of the empire, where they created a sense of participation in this imperial event.[90] Most of them are now lost, but among those which have survived is *Queen Victoria's Diamond Jubilee*, which was produced by Paul's Animatograph Works and seems to be one of the most complete representations of the Jubilee procession.[91] In 1936 R.W. Paul spoke about the filming of the Diamond Jubilee at the British Kinematograph Society, and described the magnitude of the event and the competition among filmmakers:

> An outstanding event of 1897 was the Diamond Jubilee of Queen Victoria, with its magnificent pageantry of royalties and troops from all over the world, and the touching ceremony at the steps of St Paul's Cathedral. Large sums were paid for suitable camera positions, several of which were secured for my operators. I myself operated a camera perched on a narrow ledge in the Churchyard.[92]

90 John Barnes, *The Beginning of Cinema in England, 1894–1901. Jubilee Year 1897* (London: Bishopsgate Press, 1983), 178–199.

91 Other Jubilee films available at the BFI are those produced by the British Cinématograph Co., the Lumière firm, Prestwich Manufacturing Company, Velograph Syndicate, the Photographic Association.

92 Robert W. Paul, "Before 1910: Kinematograph Experiences", *Proceedings of the British Kinematograph Society*, no. 38 (1936), 4–5.

Paul was aware that a number of filmmakers were filming the procession and his employment of a number of cameras to film the event from several points of view aimed at creating one of the most completed and competitive views of the event. The Jubilee was a major attraction, and the middle-class entrepreneurial spirit of early filmmaking saw in this event a profitable occasion. The competition was fierce and, in order to gain as much as possible from the Jubilee, filmmakers had to offer a high-quality representation. The most established filmmakers, including Paul, employed all their technical equipment to fight competitors and present as many Jubilee scenes as possible. Paul had three cameras placed in the key places of the procession: one was situated at the corner of York Road, Westminster, and filmed the parade from a high angle; the other two on the south and north sides of St Paul's Churchyard.[93] This resulted in a series of scenes of the procession taken by Paul and two cameramen. Nine scenes from these films have survived and here we can see the Life Guards and the Bluejackets, mounted Australian and South African troops, the Canadian Hussars, the Cape Mounted Rifles led by Lord Roberts and the Naval Cavalry, the Dragoons and the Indian Escort, foreign royals, dignitaries and, of course, Queen Victoria and the Royal Family.[94] This large number of exotic troops from the colonies and local rulers taking part in the procession contributed to making Victoria's Diamond Jubilee a celebration of the empire.

The presence of individuals from colonized territories was a reminder of the power of the British Empire and thus contributed to the imperial make-up of London's ceremonial spaces. R.W. Paul's Jubilee films exemplify this transformation. On one hand, the Queen's Guard standing between the masses and the actual parades contributed to the definition of an orderly crowd observing the passage of troops, the royals and Queen Victoria. On the other hand, the sight of colonized people marching in an orderly way through a space marked by lined Guards exemplifies the domestication of

93 Barnes, 178–199.

94 Titles of the scenes: "Head of the Procession Including Bluejackets", "Head of the Colonial Procession, Canadians etc.", "Royal Carriages Passing Westminster", "Cape Mounted Riflemen Passing St Paul's", "Dragoons Passing St Paul's", "Royal Carriage Arriving at St Paul's", "Royal Princes in St Paul's Churchyard", "Queen's Carriage and Indian Escort Arriving at St Paul's", and "Life Guards and Princes North of St Paul's".

otherness in the form of troops from exotic territories, and reinforces an imperialist interpretation of the Jubilee procession. People in the crowd were able to recognize uniforms and flags, as detailed descriptions of all the troops on display were distributed during the parade. Newspapers and magazines contributed to people's knowledge of imperial uniforms and troops. On 26 June, *The Illustrated London News* published on its cover an illustration by Amédée Forestier representing Victoria during the Jubilee procession in front of St Paul's Cathedral: she is surrounded by colonial troops wearing all sorts of exotic uniforms. A two-page spread containing eighteen illustrations of "Colonial Troops in England for the Queen's Diamond Jubilee" was printed in the same issue of *The Illustrated London News*.[95] Two weeks earlier, *The Illustrated London News* published several pictures of colonial troops which "invaded" London during the celebrations for the Queen's Jubilee: "The Sierra Leone Frontier Police", "British North Borneo Police", "British Guiana Police", "New South Wales Lancers", and "The Trinidad Police."[96] Paul's Jubilee films contributed as much as *The Illustrated London News* to the creation of a colonial imagery in relation to the colonial procession, and were screened in several of London's music halls and theatres from the evening of 25 June. By August the films of the Jubilee had reached Bombay and Calcutta.[97]

Other events organized and filmed in London celebrated, at different levels, the imperial theme. From 1897, early filmmakers filmed the most important official events taking place in the capital. As Michael Chanan argues, these films display a spectacle of state which was always structured around central figures of political and military authority or symbolic leadership.[98] An ideologically-imposed consciousness also pervaded events like the Lord Mayor's Show, which was filmed in 1897 and 1909.[99] In 1898 the funeral procession of Rt Hon. W.E. Gladstone was filmed, among other

95 *The Illustrated London News*, No. 3036, Vol. CX (Saturday, June 26, 1897), 850–851.
96 *The Illustrated London News*, No. 3034, Vol. CX (Saturday, June 12, 1897), 807–808.
97 John Barnes, 195–199.
98 Chanan, 238.
99 Titles: *Lord Mayor's Show* (Chard's Vitagraph, 1897) and *The Lord Mayor's Show* (Gaumont, 1909). Both films are now lost.

companies, by R.W. Paul for Paul's Animatograph, Cecil Hepworth for the Warwick Trading Company, and George A. Smith.[100] *Heroes of Ladysmith Marching through London* was produced in 1900 by the Warwick Trading Company and celebrated the return of the troops engaged in the Siege of Ladysmith during the Second Boer War. In the same year, R.W. Paul filmed *Sirdar's Reception at Guildhall* which records the visit of Lord Kitchener, the British Commander-in-Chief of the Egyptian Army, to the Guildhall for the presentation of the Freedom of the City of London. Outside London, Victoria's funeral procession was filmed by the major British and foreign firms on Saturday 2 February 1901.[101] The manager of Charing Cross's Palace Theatre produced a large amount of footage representing this event from six vantage points from Cowes to Windsor, and exhibited his film on Monday at the Palace Theatre.[102] Official visits of foreign heads of state, such as those of French presidents Émile Loubet and Armand Fallières in 1903 and 1908 to the Guildhall respectively, were also filmed.[103]

The coronation of King Edward VII, another major event which marked the heyday of the British Empire, took place in 1902 and also became the subject of several topical films.[104] Edward VII and Queen Alexandra were crowned at Westminster Abbey on 9 August 1902, and the parades preceding the event were filmed by several filmmakers, including Cecil Hepworth, R.W. Paul and William Dickson. The coronation of Edward VII and its filming offered early filmmakers a chance to make successful films. It actually offered them two opportunities: in June 1902, London looked forward to seeing Edward's ascent to the throne, and several filmmakers

100 Titles: *Gladstone's Funeral* (Paul's Animatograph, R.W. Paul, 1898), *Funeral of Mr Gladstone* (G.A. Smith, 1898) and *Gladstone's Funeral* (Warwick Trading Company, Cecil Hepworth, 1898).

101 See Gifford, 65–68.

102 Andrew Horrall, *Popular Culture in London c. 1890–1918: the Transformation of Entertainment* (Manchester and New York: Manchester University Press, 2001), 45.

103 Titles: *President Loubet's Visit to London* (Warwick Trading Company, Will Baker, 1903) and *Visit of President Falliès* (Charles Urban, 1908).

104 Two other events saw the presence of a great number of cameramen on the streets of London: the funeral of Edward VII and the coronation of George V. See Gifford, 162–163 and 172–177.

were there to film the event. The coronation was originally scheduled for 26 June, but two days earlier Edward VII developed appendicitis and the event was postponed. Nevertheless, on 26 June, Cecil Hepworth decided to film the streets of Westminster anyway. The result is *Disappointed London*, later introduced by Hepworth himself with these words: "Londoners, disappointed of their coronation, tried to console themselves by promenading the decorated streets."[105] In *Disappointed London* Hepworth filmed the area around Westminster Abbey, where the coronation should have taken place, with decorations welcoming the new King. In this film, the streets are indeed crowded with disappointed Londoners: carriages, coaches and omnibuses had taken over the sites where the procession should have taken place. The public which should have been contained by the King's Guards is seen in *Disappointed London* as it occupies the space staged for the exhibition of the new King and the royals. The physical separation between the space of the parade and the space of the spectator loses its shape in a film where we can observe everyday mobility on a ceremonial ground. Westminster's monumental spaces, including the Houses of Parliament, Westminster Bridge and Trafalgar Square, feature in the film and, in particular, Hepworth focuses on the Canada Arch, which had been erected on the ceremonial route from Buckingham Palace to Westminster Abbey. Fifty-six feet high, sixty feet wide and twenty-five feet wide, and decorated with wheat sheaves from Manitoba, and Canada's national emblem, the maple leaf, it was intended to boast imperial sentiments. It was lit with electric lights and messages on both sides of the arch were visible night and day: "Canada. Britain's Granary. God Bless Our King and Queen" and "Canada. Free Homes for Millions. God Bless the Royal Family." Canada Arch, with its roof of crown formation, also features in other coronation films made after Edward VII recovered from his appendicitis, for example in James Williamson's *Coronation Through London*. The arch was indeed a novelty and, at the same time, a new temporary imperial landmark in the urban geography of London.

105 Hepworth added this text to the original film in the 1920s in a compilation called "Through Three Reigns".

Later that summer, Hepworth succeeded in filming the actual coronation and his films were particularly successful. In his autobiography he wrote:

> Our success with the Coronation seems to have inspired a spate of news-realism, what with Lord Kitchener at Ipswich, the procession of the King and the Queen around London in October, the arrival of the German Emperor, Joseph Chamberlain's departure for South Africa, the state opening of the Parliament in February of the following year, 1903, and the launch of the third *Shamrock*.[106]

Events like the coronation contributed to the creation of a mood of imperialist determination and it was filmed by pioneer filmmakers with the usual attention to the sights and landmarks of the empire in central London, whether these were temporary (like Canada Arch) or permanent (like the familiar view of the Big Ben Tower or Westminster Abbey).

As the analysis of *Old London Street Scenes* has shown, everyday life and the built environment in the central districts of London revealed the imperial character of the city. The imperial message, as we have seen here, emerges even more vigorously when the central districts of London are filmed during events such as the Diamond Jubilee or the coronation of Edward VII. Exceptions to the dominant image of the city as the nexus of the empire are to be found in films of popular districts suggesting the presence of an unruly urban space characterized by disorder and the unchecked movement of people and things. This is the case of *Petticoat Lane* and other, rare, films of popular markets and districts. As Jerry White suggests, the exotic mix of people from abroad working and shopping in Petticoat Lane Market was a symbol of multicultural life in London.[107] On the other hand, films of processions contain images of orderly and regular movement on the stage given by the streets of London: masses of citizens watching the parades formed an essentially Western, multi-class crowd observing colonial troops and cultural otherness in the form of display. Films thus reveal different expressions of a binary opposition between British and foreign

106 Cecil Hepworth, *Came the Dawn* (London: Phoenix House, 1951), 63.
107 White, 128–129.

subjects. What films like *Old London Street Scenes*, *Petticoat Lane* and *Head of the Colonial Procession* have in common is an attention to moving material in the city. As we shall see in the following chapter, literal and figurative motifs of movement in early films of London reveal modern practices of the city concerned with both new perceptions of mobility expressed by developments in urban transport and with the display of artefacts, people and things from the colonies.

CHAPTER 2

Connecting Places: Film, Transport and Display

In which we visit one of the most extraordinary buildings of the nineteenth century and encounter the most exotic examples of the animal kingdom in the heart of London.

Between the beginning of the nineteenth century and the Great War the expansion of direct colonial dominion influenced the geographical imagination of the British. Colonial exhibitions and world fairs displayed peoples, animals, places, cultures and products which were underpinned by the technological modernity offered by new constructional, industrial, communication and transport technologies. An article published in 1862 in the *Illustrated Weekly News* proclaimed "this is the age of exhibitions."[1] Exhibitions were also destined to become a favourite subject of early actuality films of the city. As a product of the forces of Victorian capitalism, early films contributed to urban cultures of display which, as Antoinette Burton argues, led to a re-configuration of metropolitan culture enforced by commodity capitalism for the display of colonial subjects and things at world exhibitions.[2]

Films of London exemplify forms of literal and figurative movement in, to and from the city. The first part of this chapter investigates the relationship between early films of London and exhibitions, including newly-built museums, the Crystal Palace and the London Zoo. In particular, I discuss

1 *Illustrated Weekly News*, October 12, 1862: 2.

2 Antoinette Burton, "London and Paris through Indian spectacles. Making a spectacle of Empire: Indian travellers in fin-de-siecle, London", *History Workshop Journal*, no. 42 (1996): 128.

the exhibition of exotic or colonial subjects and objects and the display of technological novelties. In a discussion on the role of cartography in forging spatial imagination, Denis Cosgrove argued that during the final years of the nineteenth century, the global imagination of many Europeans was dominated by a pervasive sense that spatial limits had been reached because, over the course of the century, technological inventions brought geographically distant regions into regular contact. The conception of the world as exhibition, Cosgrove suggests, found its expression in the nineteenth century in the colonial exhibition or world's fair, and became fundamental to a modern Western geographical imagination.[3] Geographical imagination was also defined by the ways in which modernity and the past offered a binary opposition which defined the British consciousness of modernity itself. The categories of the modern were perceived as belonging to Britain and its colonial explorers, who brought into the country the "others", in the form of people, things, photographs and moving pictures. In contrast, the colonized people were often perceived as non-modern, ancient, primitive, traditional and superstitious. Colonial imperialism thus defined Britain's modernity by establishing clear boundaries between the colonized and the colonizer. As Thomas Richards suggests, the ruling class in late Victorian Britain accumulated knowledge in cultural and scientific fields which was enlisted in the service of the empire and led to the consolidation of a corpus of global knowledge and power.[4] Public display was a fundamental part of imperial knowledge as it aimed at containing the world as exhibition. Early cinematography contributed to this project by containing exhibition spaces in an artificially-confined mass-public viewing environment.

Siegfried Kracauer identified the tendency of the camera in early cinema to remain fixed to the ground. During the early days of the cinematograph,

3 Cosgrove, 220–224. See also Paul Greenhalgh, *Ephemeral Vistas: History of the Expositions Universelles, Great Exhibitions and World's Fairs* (Manchester and New York: Manchester University Press, 1988).

4 Thomas Richards, *The Imperial Archive: Knowledge and the Fantasy of Empire* (London and New York: Verso, 1993), 73–77. On exhibitions and empire in the nineteenth century see Peter Hoffenberg, *An Empire on Display* (Berkeley: University of California Press, 2001).

as Kracauer suggests, this tendency encouraged pioneer filmmakers to concentrate on objects in movement.[5] In the second part of this chapter, the discussion shifts towards the relationship between cinema and urban movement as it literally emerges in the images of movement across the suburban spaces in films representing and exploiting forms of old and new means of transport. Less concerned than other films with the imperial *façade* of London, these films reveal the roadway as a place of increasingly free movement about the city due to the development of tram systems. Trains, carts, bicycles, electric-, steam- and horse-drawn tramcars were often the subject of early actuality films; the camera was also brought onto trains and trams in order to film the experience of movement, in the so-called phantom-ride films. Technological innovations contributed to altering the perception of global distances, and also caused changes to the experience of urban space. Transport and the various types of display in the Victorian and the Edwardian eras contributed to connecting disparate geographical areas across the empire as much as inner districts of the capital and its suburbs. In what follows, I investigate different ways in which centre and periphery were connected by non-fiction films both in relation to the connection between London and the colonies, and that between the central districts of the city and the suburbs.

Exhibitions on film: Displaying the empire in London

As Edward Said suggests, histories of European modernism often leave out the great influences of non-European cultures on the western metropolis.[6] Timothy Mitchell and Felix Driver are among a small number of scholars who have looked at these influences. Mitchell has argued that modern

5 Siegfried Kracauer, *Nature of Film: The Redemption of Physical Reality* (London: Denis Dobson, 1961), 34.
6 Said, *Culture and Imperialism*, 292.

ways of seeing the world were exemplified by the growth of museums and exhibitions throughout the nineteenth century, and by a political exploitation of these sites which was aimed at displaying national, cultural and racial differences. Exhibitions, Mitchell continues, were strictly bound to ideas of improvement and progress, and a reflection and production of that sense of scientific and political certainty which aimed at rendering "history, progress, culture and empire in objective form."[7] Exhibitions enabled Europeans to apprehend and "colonize" reality by offering an orderly and organized experience of the world as though it was a picture. Travel photography and painting, catalogues and guidebooks all contributed to the definition of a picture of domesticated otherness; films representing colonial exhibitions, as well as distant and exotic destinations, achieved the same result.

Victorian Britain also saw the emergence of a modern culture of exploration which, as Felix Driver suggests, was related to empirical and geographical knowledge and required a wide range of material and imaginative resources. The display of newly discovered worlds in London's exhibitions and museums contributed, as Driver argues, to the development of an imperial scientific project, and to the metropolitan experience of the empire, by representing the identities of the colonized and the colonizer and the boundaries which defined them.[8] The work of pioneer filmmakers also reflected the emergence of an imperial culture of exploration. Their films resulted in a visual map of London characterized by the impact of imperial cultures of exploration and technological progress on the urban space. Tom Gunning has explained that nineteenth-century exhibitions trained audiences to take part in a commodity-based visual culture which created the right conditions for ethnographic representations for the masses.[9] Gunning's argument echoes Walter Benjamin's, who defined world

7 Timothy Mitchell, *Colonizing Egypt* (Berkeley and Los Angeles: California University Press, 1991), 2–19, 160.

8 See Felix Driver, *Geography Militant: Cultures of Exploration and Empire* (Oxford and Malden, MA, 2001), 1–48 and 167–169.

9 Tom Gunning, "The World as Object Lesson: Cinema Audiences, Visual Culture, and the St Louis World's Fair, 1904", *Film History*, vol. 6, no. 4 (1994): 422–44.

exhibitions as "places of pilgrimage to the fetish commodity."[10] Films of exhibitions reflected the representational strategies of colonial and technological displays. The visual power of exhibitions in London led several companies to film these places, and to offer a cinematic representation of spaces which were themselves a representation of colonized cultures and technological development. Filmmakers were induced by the popularity of these places to exploit their image and offer it to those who were unable to visit them in person, and to those who had visited the exhibitions and wanted to re-experience their attractions. In 1911, a film company called New Agency produced a film, now lost, representing two of the major attractions of London: *The Crystal Palace and a Glimpse of the Zoo*. As we shall see, the Crystal Palace and the London Zoo were among the most popular exhibitions in London and had been filmed several times in the previous fifteen years.

The London Zoological Gardens

Opened in 1828 and situated at the northern edge of Regent's Park, the location of the Zoological Gardens was particularly meaningful. Created by the architect John Nash at the beginning of the nineteenth century, the project of Regent's Park was based, as Richard Sennett suggests, "on the principle of park-as-lung, but adapted to a city where greater speed was possible."[11] Sennett has explained how Nash's plan worked against a socially connected use of the park by creating a rapidly-moving busy road around the park aimed at discouraging the organization of groups on its open space.[12] The construction of Regent's Park embodied the development of urban modernization by enabling the movement of a great number of individuals and disabling the movements of groups understood as an

10 Walter Benjamin, "Paris: Capital of the Nineteenth Century", *Perspecta*, vol. 12 (1969), 167.

11 Richard Sennett, *Flesh and Stone: The Body of the City in Western Civilization* (London and Boston, MA: Faber and Faber, 1994), 325.

12 Sennett, 325–329.

agglomeration of men and women presenting very different characteristics from those of the individuals composing the crowd. The opening of the London Zoo on this space resulted in a direct link to imperial culture. Indeed, the nineteenth-century development of zoological gardens in Europe has been discussed in terms of imperial showcase created by the ruling class ostensibly to display their colonial empires.[13] Nash's original plan thus acquired a space where, as Jonathan Schneer argues, a Londoner's identity as resident of an imperial metropolis was fully confirmed.[14] The Zoological Society's project to list and classify all species of animals was inherently imperialist, and capturing and caging wild beasts from every corner of the empire and displaying them in London was a figuration of Britain's imperial enterprise.

Until 1846 entrance to the gardens was restricted to an elite membership, but at the turn of the twentieth century the Zoo became a much-loved form of popular recreation: imperial culture, as it was expressed in the Zoo, was underpinned by scientific progress and it attracted popular interest. Books, articles, and music-hall songs were devoted to the animals kept in captivity in Regent's Park: "Birth of Kangaroo at the Zoological Gardens" titled *The Illustrated London News* in July 1896.[15] The song *Walking in the Zoo is the OK Thing to Do* was written by Alfred Vance in 1870 and at the turn of the century was still sung in London's music halls. At the same time, controversies on the zoo's treatment of animals constantly appeared in the newspapers.[16]

The Zoo hosted animals from all over the world. The most impressive and popular subjects of the animal kingdom, such as elephants and tigers, represented the colonized territories. Films representing these animals were thus a further reminder of the submission of vast territories to the British Empire. When Alexandre Promio arrived in London in 1896, he not only filmed architectural landmarks like Tower Bridge, the Houses of Parliament

13 See, for example, Eric Baratay and Elisabeth Hardouin-Fugier, *Zoo: a History of Zoological Gardens in the West* (London: Reaktion Books, 2003).

14 Schneer, 97–98.

15 *The Illustrated London News*, No. 2985, Vol. CIX (Saturday, July 4, 1896): 345.

16 Schneer, 98.

and a number of street scenes. The Lumière cameraman also made three short films in the London Zoological Gardens: *Lions*, *Pelicans*, and *Tigers*. Promio was probably the first filmmaker to film the beasts of the London Zoo. During the following fifteen years many British film companies sent their cameramen to Regent's Park in order to film the attraction offered by exotic animals. The result is a group of films documenting life in the zoo, and focusing on its attractions and, occasionally, its visitors. *London: Zoological Gardens* was produced by Chard & Co in 1897, and included "Feeding the Pelicans" and "Polar Bears." The Warwick Trading Company produced *Feeding the Tigers* in 1899 and Charles Urban made *A Ramble around the Zoo* one year later. The Walturdaw Company produced *The Zoo Series* and *A Peep at the Zoo* in 1905, and five years later released *A Peep around the Zoo*. British Pathé produced *At the Zoological Gardens* in 1909, and Gaumont made *A Visit to the Zoological Gardens* in the same year. In 1911, Gaumont filmed *Who's Who at the Zoo*, and one year later released *A Visit to the Zoological Gardens: the Ruminants*. In 1912 Kineto produced *Snapshots of the Zoo*, and one year later B&C filmed *A Ramble through the Zoo*. Two years later Wardour produced *A Visit to the London Zoo*. Most of these films are now lost. Among the few which have survived, *Scenes at the London Zoo* is arguably the most remarkable for the variety of shots and subjects. Filmed by Cecil Hepworth for the Warwick Trading Company in 1898, it offers a series of shots representing a man playing with snakes, a big bear, a llama, a camel, a dromedary, several flamingos, pelicans, a lion, parrots, and toucans. Most strikingly, Hepworth's *Scenes at the London Zoo* shows, in its two longest sequences, children and adults riding an Indian elephant in the Zoo's alleyways.[17] Iconic representations of elephants in the nineteenth century were a reminder of the British imperial adventure in India: for example, the elephant symbolizes Asia in one of the allegorical sculpture groups at the four corners of the Albert Memorial in Kensington Gardens. Elephants at the London Zoo had the same purpose: they were the most imposing animal in the Regent's Park Zoo, and those which best

17 *Elephants at the Zoo* was also the title of an 1898 film produced by Mutoscope and Biograph and filmed by William Dickson.

represented the most exotic territories of the British Empire. The docile elephant was devoted to the amusement of children and adults, and well exemplified the Indian subordination to Britain. The cinematic image of this symbol of submission served to spread the imperial message throughout the country, for the benefit of people who were unable to visit London and its attractions. Similarly, Hepworth filmed Royal Bengal Tigers at the London Zoo, another symbol of the submission of India to Britain.

The Crystal Palace

The Crystal Palace in Sydenham was also very popular with filmmakers and audiences. Nicky Levell has discussed the representations of the distant Orient at the Crystal Palace, and argued that the development in transport and media in Victorian Britain enabled the more people to travel to urban exhibition sites and observe other cultures through the medium of display.[18] The Palace was a complex space and oriental crafts were also displayed as a model for European artists. Originally erected in Hyde Park to house the Great Exhibition of 1851, the Crystal Palace was moved to Sydenham in 1854 and, as Gyan Prakash explains, it displayed the significant roles of classification and naming, and their link to imperial culture.[19] In these premises, the display of exotic civilizations contributed, according to Andrew Hassam, to offering a self-congratulatory image of Victorian Britain of which the Crystal Palace became a symbol.[20] In *All That Is Solid Melts into Air*, Marshall Berman defines the Crystal Palace as "the most visionary and adventurous building of the whole nineteenth century" and describes it as a key architectural landmark in the topography of London:

18 Nicky Levell, *Oriental Visions: Exhibitions, Travel, and Collecting in the Victorian Age* (London: Horniman Museum and Gardens, 2000), 11.

19 Gyan Prakash, *Another Reason: Science and the Imagination of Modern India* (Princeton, NJ: Princeton University Press, 1999), 22–23.

20 Andrew Hassam, "Portable iron structures and uncertain colonial spaces at the Sydenham Crystal Palace", in *Imperial Cities: Landscape, Display and Identity*, 180.

> No building in modern times, up to that point, seems to have had the Crystal Palace's capacity to excite people. As for foreigners, the Palace, more than anything else in London, became the sight they wanted to see first. Contemporary journalists reported that it was London's most cosmopolitan zone, crowded at any given time with Americans, Frenchmen, Germans, Russians, even Chinese and Japanese.[21]

Berman identifies the Crystal Palace as one of those modern environments "that promises adventure, power, joy, growth, transformation of ourselves and the world."[22] A place like Crystal Palace, and the attraction that it offered, fitted perfectly the kind of visual material sought by early film companies. The Palace offered a cosmopolitan space, rich in innovation and signs of modernity, and films inevitably depicted its attractions on several occasions.

The Crystal Palace and its display of marvels of the Victorian and Edwardian age in South London was a significant cinematic site. In 1896 Promio filmed Javanese performers in the park grounds of the Crystal Palace: *Danse Javanaise* represents a group of Javanese dancers and musicians performing in traditional costumes. In *Jongleur Javanais* a Javanese man performs a series of juggling routines with a ball, which he kicks up and balances on his shoulders and feet. Javanese musicians seated on the ground accompany his performance. The same group of Javanese performers appears in *Lutteurs Javanais*, where two men wrestle, while the juggler performs accompanied by the musicians seen in the other films. In 1898 the ascent of a gas balloon at the Crystal Palace was filmed by a cameraman of a company called Interchangeable Ltd.[23] In 1899 a Warwick Trading Company cameramen filmed *Horse Parade at Crystal Palace*. In 1900, George A. Smith made for the Warwick Trading Company a number of films at the Crystal Palace documenting its attractions: horse parades, bicycle races, and the series of sculptures of dinosaurs and extinct mammals located in

21 Marshall Berman, *All That is Solid Melts into Air: The Experience of Modernity* (New York and London: Penguin, 1988), 237–238.

22 Berman, 15.

23 Title: *Balloon Ascent at Crystal Palace*.

the park.[24] In 1903 H.M. Lomas filmed for Charles Urban *Crystal Palace Panoramas*, *Looping the Loop*, *Shooting the Chutes* and *Panorama of Palm Green*, a series of views of the Palace, its park and terraces. In 1906, Smith went back to the Crystal Palace and filmed *Togo's Heroes at the Crystal Palace* and *International Cycling Race at Crystal Palace* for Charles Urban. In 1911, both Will Barker and John Y. Brown filmed for the London Cinematograph Co. the Crystal Palace during the Pageant of the Empire held in London to celebrate the coronation of George V.[25] The Pageant, as Deborah Ryan suggests, was both an expression of imperial might and of civic pride, for it aimed at telling the history of London in forty staged scenes depicting the city as the capital of the empire.[26] The filming of this event and its attractions offered the pleasures of the Crystal Palace to the masses unable to visit in person, and gave them an image of London which was saturated with signs and sights of the opulence and greatness of the imperial metropolis, centre of a world-wide empire and of the history of Britain.

In 1902 the Crystal Palace Company organized an exhibition of motor-cars, motor-cycles and accessories, and offered a number of awards to exhibitors.[27] The Lanchester Engine Company, based in Armourer Mills near Birmingham, gave a demonstration of its motor cars at the Crystal Palace, where several cars climbed steps in the park. At least two films documenting the event have survived: *Lanchester Car Test on steps at Crystal Palace* and *Motor Car Climbing Contest at the Crystal Palace*.[28] In the late 1890s, an automobile craze swept London and several performers used motor-cars during their shows on the streets and even in music halls. As Andrew Horrall suggests, automobiles brought together "utilitarian transportation and dandified public spectacle."[29] Motor cars became a popular attraction

24 These are the titles: *Panorama View of Crystal Palace Terraces*, *Panorama of Crystal Palace Lake*, *the Swings at the Crystal Palace*, *The Prehistoric Monsters at Crystal Palace* and *Bicycle Race at the Crystal Palace*.

25 Titles: *The Pageant of the Empire* and *The Pageant of London*.

26 Deborah S. Ryan, "Staging the Imperial City: The Pageant of London, 1911", in *Imperial Cities: Landscape, Display and Identity*, 117–135.

27 *The Times* (Saturday, 15 February 1902): 14.

28 There are no credits available for these films.

29 Horrall, 67.

which did not fail to attract filmmakers: Cecil Hepworth made *Explosion of a Motor Car* (1901), one of the earliest motor-car films. Films of automobiles made at the Crystal Palace, such as William Dickson's *Motor Car Race in Regent's Park* (1900), document the popular enthusiasm surrounding the technological progress embodied by the motor car, and contributed to what Thomas Richards described as the transformation of collective and private life into a space for the spectacular exhibition of commodities.[30] Cinema and car industries were the result of nineteenth-century technological developments, and they grew together in the twentieth century. From the very early years of cinematography, cars were both the subject of films and were also used – as we shall see later in this chapter – to film the streets of the city from a moving point of view in the so-called phantom rides.

The Zoo and the Crystal Palace were two of many exhibitions and museums explored by pioneer filmmakers. The function of early non-fiction films and other colonial displays often intersected. Museums, as Barbara Black suggests, helped to legitimate the power of Britain across the vast empire by housing the spoils of the colonial project.[31] Both films of distant parts of the empire and films of exhibitions in London revealed a contribution to the accumulation of knowledge in the age of empire.

The South Kensington Museum was established in 1852 following the success of the Great Exhibition of 1851 and displayed examples of the arts and scientific achievements. Established with a similar purpose the International Universal Exhibition at Earl's Court opened in 1898. These exhibitions were also the subject of a great number of actuality films and contributed to enforcing the scientific message of imperialism in popular venues. On the 17 May 1900, Queen Victoria laid the foundation stone at the main entrance of the new Aston Webb building at the South Kensington Museum, and it was during this ceremony that the name of the museum was changed from the South Kensington Museum to the Victoria and Albert Museum. This event, which was destined to be Queen

30 Richards, *The Commodity Culture of Victorian England: Advertising and Spectacle, 1851–1914*, 72.

31 Barbara J. Black, *On Exhibit: Victorians and Their Museums* (Charlottesville and London: University Press of Virginia, 2000), 11.

Victoria's last official public appearance, was filmed by both the Warwick Trading Company and Biograph Syndicate.[32] Works were completed in 1909 and the opening of the new museum was filmed by Charles Urban in *London's New Museum*.[33] Tim Barringer suggested that the museum aimed to open up the economic and cultural possibilities of empire to the public, the inhabitants of the largest British cities.[34] Early actuality films were also aimed at the fulfilment of the imperial message; their association with urban culture of display can thus be seen as a relationship embodied in the imperial educative rule.

The inauguration of the Universal Exhibition at Earl's Court was filmed by Chard's Vitagraph in *Opening Ceremony at International Universal Exhibition*. In 1907, two actualities representing the South African Exhibition were produced by the Warwick Trading Company and Walturdaw: *Opening of the South African Exhibition* and *South African Products Exhibition*. These films, which had been made five years after the end of the second Boer War, offered images of South African artefacts and emphasized the submission of the Republic of the Orange Free State and the South African Republic to Britain. They promoted the imperial message as much as actuality films shot in the war zone itself or by fake Boer War films, and served the purpose of celebrating the might of the British Empire. Ultimately, the impact of imperialism and colonial representations on London's urban landscape emerges from these films and, as Driver and Gilbert put it, "it encourages us to consider the ways in which the experience of modernity in Europe reflected or represented the European encounter with the world beyond."[35]

32 Titles: *Her Majesty the Queen Arriving at South Kensington on the Occasion of the Laying of The Foundation Stone of The Victoria And Albert Museum* and *Laying of The Foundation Stone of The Victoria And Albert Museum*.

33 These films have not survived and their loss is particularly unfortunate as they represented a place which was important to the imperialist idea of the world as an exhibition.

34 Tim Barringer, "The South Kensington Museum and the colonial project", in *Colonialism and the Object: Empire, Material Culture and the Museum*, eds Tim Barringer and Tom Flynn (London and New York: Routledge, 1998), 15.

35 Driver and Gilbert, "Heart of Empire? Landscape, Space and Performance in Imperial London", 13.

Motifs of movement emerge consistently from early films of London and reveal both the imperial dimension of the city and, as we shall see, a local dimension concerned with the theme of movement in the city. As Donald Lowe suggests, during the nineteenth century the perception of time, space and distance was changed by the developments in transport and communication technologies.[36] The rail network and the application of steam-engines to ships made long-distance journeys more affordable, and enabled the upper middle class to travel the world and encounter distant cultures. New means of transport also altered the experience of travelling form London to the rest of country and the experience of urban space in the outer districts of the city.

The railway and London's suburban spaces

Early films of London exemplify forms of literal and figurative movement and effectively contribute to the understanding of urban life with a focus on a spectacle of narrativized motion. Giuliana Bruno has argued that early cinema shared a constant reinvention of space with nineteenth-century travel culture defined by new architectural sites of transit, such as railways and exhibition halls, which embodied mobility. Bruno suggests that the previously unseen "spatio-visuality" produced by new architectural forms changed the relation between spatial perception and bodily motion, and prepared the ground for the invention of the moving image, "the very epitome of modernity."[37] Architectural forms such as arcades, railway stations, department stores, exhibition halls, glass houses and winter gardens defined the geography of modernity within the urban space, and constantly featured in its early cinematic representation.

36 Donald Lowe, *History of Bourgeois Perception* (Chicago: University of Chicago Press, 1982), 35–38.

37 Bruno, *Atlas of Emotions, Journeys in Art, Architecture, and Film*, 17–21.

Early filmmakers were attracted to these sites, both in the inner districts of the city and the suburbs. While most studies of the imperial city focus on the city centres, the work of Felix Driver and David Gilbert on imperial London has attempted to widen the conventional focus on the ceremonial core at Westminster and, by considering different ways in which London can be conceived as an imperial city, to map various alternative sites in the landscape of the imperial city, including the docklands and the suburbs.[38] As Driver and Gilbert have argued:

> Imperial sentiment [...] was as much about the suburbs as the centre, about ordinary lives as much as patriotic spectacles, about mothers and children as much as soldiers and politicians.

Driver and Gilbert have discussed the representation and negotiation of cultures of imperialism in the urban landscapes of London, and are among the few scholars who have considered the role of the suburbs in the relationship between the city and the circulations of empire.[39] Suburban London was defined by the modernization of transport and consequently by a different spatial relationship between the inner city and its suburbs.[40] A small group of early films documented life in London's suburbs, and can contribute to our understanding of their role in shaping the experience of the city. These films reveal practices of movement about the city and are less directly concerned with imperial iconography than the films discussed so far. In what follows, I shall address this alternative image of the imperial city by investigating the depiction of transport and movement outside Westminster and the City.

The suburban and urban experience in the city can be investigated by taking into account the technological development which shaped suburban

38 Driver and Gilbert, "Heart of Empire? Landscape, Space and Performance in Imperial London", 11–28.

39 Driver and Gilbert, "Heart of Empire? Landscape, Space and Performance in Imperial London", 13–14.

40 Francis M. Thompson, "The Rise of Suburbia", in *The Victorian City: a Reader in British Urban History: 1820–1914*, eds Robert J. Morris and Robert Rodger (London and New York: Longman, 1993), 149–151.

life, in particular new forms of urban transport such as the electric tramcar. The rail network, on the other hand, shaped the experience of travelling from a city to another. In the nineteenth century, the railroad contributed to the development of spatial perception, a process which, as Wolfgang Schivelbusch argues, was peculiar to modern times and led to the multiplication of visual impressions for the sense of sight to deal with.[41] Lynne Kirby has extensively discussed the connection between early cinema and the railroad, and the "parallel tracks" taken by modern transport and visual culture at the beginning of the twentieth century. Films, Kirby suggests, arrived at the end of a century defined mechanically and temporally by the railway. Both technologies caused fears about the eradication of tradition and concerns about moral and social corruption. Kirby also points out how the railway affected both the interurban and the urban landscape on a global scale, as its stations became defining places of the new urban culture and economy.[42] The economic relation between the city and the rest of the country was marked by the railway station, the place where the entry into and the exit from the city was organized and where a scale version of city marketplaces developed. During the second half of the nineteenth century, as railway travel enabled the middle class to escape the city for vacations, a large number of railway stations were constructed in London. Nevertheless, in the 1850s the railway did not make an immediate impact upon the city and commuting to London rose significantly only at the end of the century. Until the 1880s, many working-class individuals never travelled by train.[43] King's Cross opened in 1852 and was the first of many major railway stations in the city centre to be added to the four existing ones: London Bridge (1836), Euston (1837), Fenchurch Street (1841), and Waterloo (1848). Between 1852 and the end of the century, at least ten stations opened in central London: Paddington in 1854, Victoria in 1862,

41 Wolfgang Schivelbusch, "Panoramic Travel", in *Nineteenth-Century Visual Culture*, eds Vanessa R. Schwartz and Jeannene M. Przyblyski (New York and London: Routledge, 2004), 94.

42 Lynne Kirby, *Parallel Tracks: the Railroad and Silent Cinema* (Exeter: University of Exeter Press, 1997), 6–10, 135.

43 See Porter, *London: a Social History*, 273–275.

Farringdon in 1863, Charing Cross in 1864, Moorgate in 1865, Cannon Street in 1866, St Pancras in 1868, Liverpool Street in 1874, Blackfriars in 1888, and Marylebone in 1899. The railway and its stations brought radical changes in city life: overcrowding in the city centre was exacerbated by railway-building, as between 1850 and 1900 over 100,000 people had their homes destroyed by the railways.[44] The railway encouraged new suburban housing for the working class. During the last decades of the century, the workmen's fare introduced by the 1884 Cheap Trains Act created a new working-class suburbanization, which brought new changes to the urban social fabric as the working class began moving to the suburbs.[45]

Railway stations represented a sign of technological modernization and had a strong impact on urban development and suburbanization, and as such they were filmed several times by pioneer filmmakers. Unfortunately most of these films are lost but their titles usefully reveal their subjects. Birt Acres filmed Finsbury Park Station in 1896, Cecil Hepworth made *Leaving Victoria Station* in 1901, and in 1904 Charles Urban produced a film shot *Around the Paddington Engine Yard*. In 1906 British Gaumont produced two films depicting Liverpool Street Station: *Liverpool Street Station Exterior* and *Liverpool Street Station Interior*. *Demolition of Old Railway Station, Croydon* documented the end of Croydon Central railway station, which closed in 1890, and was filmed in 1897 by Adolphe Langfier for Bender & Co. The loss of these films denies today's audience of a view of London terminal stations, and their vital movements of people and goods which could have been an interesting subject of analysis. Luckily, a few films representing railway traffic around London have survived.

Irish Mail (L & N W Railway – Taking up Water at Full Speed) was produced by British Mutoscope and Biograph Company in 1898 and represented the train handling the Irish Mail for the Government between Euston to Holyhead on the London and North Western Railway.[46] It

44 Ibid., 278.

45 Ibid., 279.

46 Biograph produced other two films on the same route: *Menai Bridge – The Irish Day Mail from Euston Entering the Tubular Bridge over the Menai Straits* (1898) and *Conway Castle – Panoramic View of Conway on the L & NW Railway* (1899). In

was filmed as the train passed south through Bushey Station, just outside London. The film mostly focuses on the express locomotive *Jenny Deans*. The locomotive is the central "character" of the film and the result reminds one of the later classic produced in 1946 by the GPO Film Unit, Harry Watt and Basil Wright's *Night Mail*. *Irish Mail* displays two technological developments, the high-speed train and the cinematograph, which brought a distinctive new experience of time and space. On the one hand, the steam locomotive, which was introduced during the first half of the century, had become increasingly efficient: the characteristic "blackberry black" locomotive of the London and North Western Railway gives a strong impression of power and speed. On the other hand, the technique employed to film its run is a significant technological achievement: a camera was fixed on another train running in synchrony alongside the *Jenny Deans* express train. The difference between *Irish Mail* and other films depicting trains made at the end of the ninetieth century is striking: Charles Goodwin Norton's *Railway Traffic on the London and North Western Railway* was filmed in 1897 and also represented the LNWR train traffic.[47] In this film, which records the arrival of three trains in as many shots, the camera is static and fixed on a platform of the Sudbury and Wembley station: visually the result does not present new ways to film speed and movement. On the contrary, *Irish Mail* offers an image of speed and strength which celebrates both the LNWR locomotive and the medium employed to represent it. In this film, train and filmmaking literarily run on parallel tracks displaying a technological progress which is quintessentially modern. From a visual point of view, the interaction between film and transport exemplified in a film such as *Irish Mail* influenced, as Kirby suggests, the mode of representation established by early filmmakers:

1897 Birt Acres produced a film with a similar subject: *Great Northern Mail Train Catching Post Bags* (Northen Photographic Works). A year later, R.W. Paul filmed the London Express in Wood Green.

47 In 1899 Norton filmed *London & North Western Railway Scotch Express*, which has not survived.

> As a *perceptual* paradigm, the railroad established a new, specifically modern mode of perception that the cinema absorbed naturally. The kind of perception that came to characterize the experience of the passenger on the train became that of the spectator in the cinema.[48]

Films taken from tramcars or cars in movement also had a local dimension and were filmed in the familiar streets of the city where they would have later been screened. *Panorama of London* was filmed in 1908 by Gaston Quiribet for the Hepworth Manufacturing Company, and represented a car-ride from Charing Cross to Bank. One year later Quiribet directed *The West End of London*, another car-ride where he filmed the most famous sights of the West End, such as Piccadilly Circus and Leicester Square. We have already seen how in 1909 Gaston Quiribet filmed for the Hepworth firm *City of Westminster* and *City of London*, both photographed from an Argyll car. Ten years earlier R.W. Paul had produced *On a Runaway Motor Car through Piccadilly Circus*, which was filmed from the front of a moving vehicle. In 1901, the Warwick Trading Company produced *Ride through Piccadilly Circus*, which was filmed from a carriage driving down Coventry Street. In the late Victorian and Edwardian eras, animal-, steam- and electric-powered tramcars co-existed in British cities, and the great majority of phantom-rides films were filmed from a train or from a tram in order to transport the spectator along the city's crowded streets. Giuliana Bruno has used expressions such as "view-racking" and "view-sensing" to discuss the ways in which early cinema represented urban space by reproducing the captivating motion of cities through an experience of perception offered by tram-ride films where the camera becomes the vehicle, that is a spectatorial means of transportation.[49] In London, phantom–ride films reproduce fluidity, ease and freedom of movement in the city as it was experienced by the Victorians and the Edwardians as a result of the improvement of public transport and street safety, making a case for the civic pride which characterized the imperial city.

48 Kirby, 7.
49 Bruno, 17–21.

West London: Ealing

London transport vehicles were used to produce views of the streets of the city from a perfect vantage-point. For example, *Scenes Taken from a Moving Train* (1897) was filmed in Clapham Junction and produced by British Cinematograph, while *Tramway Ride from Shepherd's Bush to Southall* (1903) was filmed in West London and produced by the Warwick Trading Company. Both films are now lost, but the choice of Clapham and Shepherd's Bush, where the first electric tram services started, reveals an interest in new forms of urban transport. These actuality films documented processes of urbanization and suburbanization through the cinematic use of London's public means of transport. In 1903 the development of an electric-powered tram system in London brought affordable public transport to the masses. The London County Council, which was created in 1889 and was responsible for traffic management, slum clearance and new house building, adopted a policy of municipal control over public transport. The Progressive Party, which led London County Council in its early years, adopted a low-fares policy aimed at relieving overcrowded accommodation in central London, and in 1903 electric trams began to run south from the Thames to Clapham and Tooting.[50] London's first private electric service was later developed in Ealing, West London. On 4 April 1901, the electric tram routes from Hammersmith to Kew Bridge, from Shepherd's Bush to Kew Bridge via Chiswick, and from Shepherd's Bush to Acton, were opened, and operated by London United Tramways. The route was later extended to Ealing and Southall, and the network was inaugurated on 10 July. As Patrick Keiller suggests, Ealing's local authority opposed the opening of the line in the belief that Ealing's exclusivity and its status as the "queen of the suburbs" would be undermined by making it easier for working-class people to commute from Ealing into London.[51]

50 Theodore Barker, "Urban Transport" in *Transport in Victorian Britain*, eds Michael J. Freeman and Derek H. Aldcroft (Manchester and New York: Manchester University Press, 1988), 157–160.

51 Patrick Keiller, "The City of the Future", in *Picture Perfect: Landscape, Place and Travel in British Cinema before 1930*, eds Laraine Porter and Bryony Dixon (Exeter:

Panorama of Ealing from a Moving Tram was filmed on Wednesday 10 July 1901 by William Kennedy-Laurie Dickson for the British Mutoscope and Biograph Company, and is one of four Biograph films representing Ealing on the day of the inauguration of the electric tram service.[52] The camera was located on the top deck of the electric tram passing the busy tree-lined Broadway, Mall and Uxbridge Road alongside Ealing Common and towards Acton Town. Exploring contrasts between the old city fabric, the past, and the present-day experience, Patrick Keiller has argued that *Panorama of Ealing from a Moving Tram* marked the celebration of two key moments in the modernization and suburbanization of West London. On the day when the electric tram service was inaugurated, Ealing also became the first local government district in Middlesex to be incorporated as a London municipal borough.[53] Therefore *Panorama of Ealing from a Moving Tram* was filmed on a very special day for the inhabitants of Ealing, and the presence of a number of flags flying from the buildings and the shops of Uxbridge Road (including the Union flag, the White Ensign and heraldic flags), which are clearly visible in the film, mark the double celebrations. In *Panorama of Ealing from a Moving Tram*, the pavements are crowded with men and women with parasols, while on the road one can see an open-top tram full of passengers, cyclists and horse-drawn traps. What emerges from this film is a wealthy and lively suburban area, and even the pleasant weather seem to contribute to the festive atmosphere in the streets of Ealing. This film reveals an image of suburban life charac-

---

the Exeter Press, 2007), 104–105. Ealing was referred to as "queen of the suburbs" because of its wide, tree-lined streets.

52 See Richard Brown and Anthony Barry, *A Victorian Film Enterprise – The History of the British Mutoscope and Biograph Company 1897–1915* (Trowbridge: Flicks Books, 1999), 295. *Panorama of Ealing from a Moving Tram* is to the only extant film. The other titles were *Distinguished Guests Leaving the Power House, The First Trams Leaving Shepherd's Bush for Southall* and *Panorama at Ealing Showing Lord Rothschild Declaring Line Open*. In 1903, Biograph also produced a series called *Opening of the New Electric Tram Route* and which included "Turnham Green", "Kew Bridge", "Apple Orchards", "Twickenham", "New Power House", "Bushey Park", "Hampton Court", and "Guests Looking Around Marquee."

53 Keiller, 106–110.

terized by orderly movement of bodies and vehicles along the streets and pavements of Ealing Common. The Metropolitan District Railway had reached Ealing in 1879 and, as Ana Vadillo has argued, facilitated the control of London's expanding boundaries which contributed to the gradual disappearance of the slums, and to the creation of suburbs of rows of semi-detached houses linked to the city centre by the railway.[54] The electric tram lines served the same purpose and contributed to resolving the problem of overcrowding in the city centre by connecting the centre to the periphery. In this film Dickson ignored the overpopulated districts of central London and focused on the experience of suburban space in the regular streets of Ealing in opposition to the unruly experience of urban life in the slums.

North London: From Seven Sisters Road to Euston Road

Early actuality films of London document life in the city suburbs: Ealing, but also Hampstead and High Barnet.[55] *Horse Drawn Traffic in Seven Sisters Road* was produced by Charles Goodwin Norton in 1898 and consists of a different view of a suburban street in North London. Norton, who shot the film with his Velograph camera, was a well-known lanternist who introduced films into his programme in 1896.[56] *Horse Drawn Traffic in Seven Sisters Road* was shot in north London, not too far from Norton's studio in St Pancras, and filmed from a moving horse-drawn tram on Seven Sisters Road running from Holloway Road north-eastward. The camera looks right and records the horse-drawn traffic towards Camden Road and Caledonian Road. A number of pedestrians and the buildings on the

54 Ana Vadillo, "Poetics on the Line: the Effect of Mass Transport in Urban Culture", in Helena Michie and Ronald R. Thomas, *Nineteenth-Century Geographies: the Transformation of Space from the Victorian Age to the American Century* (New Brunswick, NJ and London: Rutgers University Press, 2003), 237–250.

55 *Carousel: Hampstead* was probably filmed in 1900. A copy of the film is preserved at the NFTVA but no credits are available. Birt Acres filmed *Corner of Barnet Fair* and *Band Marching Down a Street* in 1896.

56 Barnes, *The Beginning of Cinema in England, 1894–1901*, 186.

right-hand side of Seven Sisters Road are also filmed. This was a suburban highway constructed in 1833 by the Metropolitan Roads Board as an extension of Camden Road, and provided a road link from Holloway to Tottenham, through Manor House and Finsbury Park. The social history of the area was entirely different from that of wealthier district of Ealing. In nineteenth-century London, suburbanization was attractive because of the dreadful habitations in the city centre. Nevertheless, suburban development had partly turned into slums and the Seven Sisters Estate, between Seven Sisters Road and the Hornsey Road, had substantially worsened since the 1850s.[57] In the film made by Charles Goodwin Norton, Seven Sisters Road appears busy with horse drawn trams, men and women. The presence of several delivery coaches on the street and people engaged in everyday activities suggests that the film was made during a week day. The difference between Ealing's festive atmosphere and the everyday activities depicted in *Horse Drawn Traffic in Seven Sisters Road* is striking and records very different suburban areas. The streets of Ealing Broadway are clean, trees stand on the side of the Mall, and people walking on the pavements are smartly dressed. A branch of the London County Bank and a Phillips music shop can be seen on the left-hand side of the road. On the contrary, no flags are displayed in Seven Sisters Road, most people here wear working clothes, no trees are visible, and the buildings are less attractive than those seen in *Panorama of Ealing from a Moving Tram*. In that film Ealing appears at its best, as the County Council had arranged decorations and the cleaning of the streets in view of the events of 10 July. The Biograph film of Ealing and Norton's representation of Seven Sisters Road offer contrasting views of suburbanization in London at the turn of the twentieth century. While Ealing emerges as a wealthy and advanced borough, Seven Sisters Road appears to be a suburban development which has yet to enter into the modernization process signalled by the electrification of the tram line. These diverse representations of life in the suburbs reveal the uneven processes of modernization and urbanization which shaped London in the nineteenth century, and which were recorded in non-fiction films.

57 Porter, *London: A Social History*, 373.

Despite the presence of modern electric tram lines in certain districts of the city, at the turn of the century horse-drawn trams were still the main form of public transport in London, at least overground, and provided Londoners with a cheap and relatively reliable transport network. In 1896 the combined omnibus and tramway traffic was 580 million passengers a year; an average horse tram, like that depicted by R.W. Paul in *Horse Drawn Single Storey Tram in a London Street* (1898), could seat up to fifty passengers, twice the number of passenger that a horse bus could seat.[58] At the end of the 1890s Charles Goodwin Norton made two other films representing the horse-drawn tram and traffic in Euston Road, which was one of the busiest areas of north London and also the place where Norton lived and worked. Norton filmed *Horse Drawn Traffic Viewed from Elevated Position* (1897) from the Midland Grand Hotel, overlooking Euston Road and Pancras Road. *Horse Drawn Traffic in Euston Road* (1898) was filmed at the at junction between Euston Road and Woburn Place, looking towards St Pancras Station and Pentonville Road, opposite the site where John O'Connor painted the famous *From Pentonville Road Looking West: Evening* in 1884. In both films, Euston Road is represented as a dynamic and busy place, with a great number of horse trams and buses, pedestrians, carriages, and coaches. Norton has captured the character of an area which by the end of the nineteenth century had become one of London's busiest joints as a result of the opening of three railway stations between 1837 and 1868. A few years later, in 1911, Bert Haldane used the nearby Euston Arch as a location for *A Burglar for One Night* in one of the earliest mature uses of London locations in fictional feature-length films. Here, a manual labourer goes to London looking for work after being laid off. Unsuccessful in his search he decides to burgle a house. Suspended between social critic and dime-novel sensationalism, Haldane's film takes the cinematic city to a new use, beyond the scope of this study.

Traffic scenes such as Norton's or Dickson's illustrated the new ease of movement of people in and around the city. This type of scenes were very common in nineteenth-century visual culture, and it is not surprising

58 Barker, 150–155.

that one of the very first moving images of London, preceding the official date of birth of the cinematograph, represented the rhythm of the city's traffic: *Traffic in King's Road, Chelsea* was photographed in 1890 by William Friese-Greene at 15 frames per second on a strip of celluloid.[59] This chapter aimed at juxtaposing these scenes, with attention to images of urban and suburban traffic as well as to films of colonial exhibitions and imperial events. The opposition between the two reveals a contradictory image of London as an increasingly complex urban space open to the empire and the world, on one hand, and an agglomeration of places where urban life could be lived locally, on the other. At the end of the nineteenth century, the development of London was characterized both by expansion and by an overcrowded city centre with traffic-choked streets, railways and masses of people at work. During the same period, the British Empire reached its heyday. Films reflected these drives and, like in Lund's painting *The Heart of the Empire*, the cinematic city is extended literally to the limits of vision and figuratively to large parts of the globe: electric tramways reaching the newest suburbs and exhibitions offering a connection to the large empire are thus spatial agents exemplifying practices of mobility in the city and the empire.

A legitimate space of empire and uneven modernity

Rather than as a hegemonic imposition from above, the imperial stance of middle-class filmmaking should be seen, as I have tried to demonstrate in this chapter, as the result of processes of negotiation and exchange among forces shaping the experience of the city in the Age of Empire. Early non-fiction films of London can also contribute to the understanding of urban modernity by focusing on the importance of self-regulation and movement

59 Gifford, 3.

in the context of late Victorian and Edwardian Britain. Throughout the nineteenth century the city was opened to view in new ways allowed by photographic realism in still photography and, at the end of the century, cinematography. Patrick Joyce argued that late Victorian photographs representing forms of transport and movement convey the contingency and fluidity of social interactions in London. The modern experience of the city, according to Joyce, was thus broken down by photography in a visual archive scrutinizing new forms of behaviour.[60] Early films conveyed the same message and their scrutiny revealed new aspects of urban behaviour. By the turn of the century, city governments consistently used photography in various activities such as sanitary and housing regulation. Unlike photography, late Victorian and Edwardian cinema was not deployed in institutions by the police, hospitals, reformatories or other city authorities. Nevertheless, the cinematic gaze scrutinizes the experience of life in the city, and reveals an image of orderly use of urban space determined by the subjective view of middle-class filmmakers. This is true for films of Queen Victoria's Diamond Jubilee, films like *Panorama of Ealing from a Moving Tram* and *Horse Drawn Traffic Viewed from Elevated Position*, and even for the busy and frenetic views of the city in *Old London Street Scenes*, where images of traffic-choked streets reveals an orderly movement of carriages and individuals.

Order and rule defined discourses of imperialism and modernity emerging from non-fiction films recording urban practices in London at the turn of the twentieth century. The focus on city life in early cinema, as Stanley Corkin suggests, is the result of the nineteenth-century relationship between positivist worldview, modernity and urbanization.[61] Corkin argued that early American filmmakers aimed at the "naturalization of their class-based perspective as reality" and, by displaying the modernity of American cityscapes, their films sought to appeal to the common desire

60 Joyce, 231.

61 Stanley Corkin, *Realism and the Birth of the Modern United States* (Athens and London: University of Georgia Press, 1996), 64.

to be aligned with the dominant culture.[62] The attempt to acculturate audiences to a modern view of the world characterized early cinema both in America and in Europe. Thomas Edison's *New York Street Scenes* (1896–1898), which included views of lower Manhattan skyline filmed from the Hudson, or Louis Lumière's *La Tour Eiffel* (1900) focused on the spectacle of urban modernity by including views of quintessentially modern icons in New York and Paris. Similarly, films of technological displays at the Crystal Palace or new forms of transport are a record of urban modernity in London. Yet, the architecture of London at the turn of the century also revealed hesitations about modernity expressed, for example, by the design of Tower Bridge. As David Gilbert and Fiona Henderson suggest, Tower Bridge rapidly came to be a powerful symbol of London throughout the world and, yet, it consisted of a "state-of-the art hydraulic machinery clothed in a Gothic shell-suit of stone."[63] The uneven modernization of London was thus reflected in films of nineteenth-century buildings that suggested a reluctance to embrace modernity, revealed by styles that drew upon Western classical architecture, such as Marble Arch or the Bank of England, or sought to revive mediaeval forms, such as the Royal Courts of Justice, the Houses of Parliament or Tower Bridge. The result is a cinematic image less unambiguously modern than that of a city like New York, with its new tall buildings dominating the skyline. As I have argued in these chapters, the specificity and meaning of early cinematic London is to be found elsewhere, that is in the imperial identity evoked by the modernity of sites like the Crystal Palace, as much as in the historicity of its public buildings. Imperialism thus characterized early films more consistently in Britain than in countries where cinematography aimed at the assimilation of a broader vision of modernity.

62 Corkin, 65–66.

63 David Gilbert and Fiona Henderson, "London and the tourist imagination" in *Imagined Londons*, ed. Pamela K. Gilbert (New York: State University of New York Press, 2002), 127.

By applying such a framework to early films of other British cities, one can find confirmation of the uneven ways in which urban modernity was displayed in early films. This relationship between the city, empire and urban modernity is not exclusively revealed in films of London. A full understanding of the cinematic city in imperial Britain requires the investigation of films of other urban experiences. The next chapter explores the ways in which the association between city and empire emerges from films of Manchester and Liverpool. This investigation reveals an urban experience where imperial iconography goes hand in hand with sign of commercial dynamism and industrial might. A continuity of themes and representational processes emerges from these films. What follows aims at re-assessing the ways in which the specific regional and urban identities of these cities result in a diverse cinematic map of the country where imperialism manifested itself as the dominant narrative of the age.

PART II

The British Isles

CHAPTER 3

Cinematic Empire: Seaside, Industry and Trade

Where further contrasts between the grandiose and the squalid are revealed and we learn that accumulating vast wealth does not protect a city from becoming ugly.

In the late 1880s, William Friese-Greene, Mortimer Evans and Frederick Varley created a camera which could take up to four or five pictures per second, too low a frequency rate for an illusion of motion but nevertheless a significant step towards the invention of the cinematograph. In 1888, Friese-Greene improved his camera and filmed a street scene in Brighton photographed at ten frames per second on oil-soaked paper.[1] At about the same time, Louis Le Prince was filming his short scene showing the traffic on Leeds Bridge. British industrial and provincial towns were indeed among the earliest subjects of films, and continued to be a recurrent landscape in non-fiction films produced in Britain until the start of the First World War. This chapter investigates the cinematic image of English cities and, in particular, draws attention to their imperial connection and iconography. If London was the political centre governing the colonies, England's industrial and mercantile cities such as Liverpool and Manchester expressed the economic aims of imperialism. Paradigms of urban life, they hold key positions on the map of turn-of-the-century Britain as places characterized by exceptional industrial dynamism and urban modernity.

1 See Dennis Gifford, *The British Film Catalogue: Volume 2, Non-Fiction Film, 1888–1994* (London and Chicago: Fitzroy Dearborn, 2001), 3.

According to Richard Sennett, the increasingly anonymous experience of social life in the great Victorian cities was given sense by their architecture and urban planning, and by a disposition of space which forged new class relationships through modes of conduct in the city.[2] Such modes and processes of self-presentation were active in the Victorian and Edwardian city, and reflected in films. The redefinition of identity and status emerging from the symbolic construction of communities was assembled in films of quotidian activities conveying a deeper set of meanings particularly concerned with an increasing ease of movement in the city. Patrick Joyce has explained how potentials for enacting freedom existed within the tension between order and movement: in nineteenth-century Manchester, for example, the exercise of freedom was invited by the process of road widening and by the construction of new streets as places of relentless passage.[3]

Tristram Hunt has drawn attention to the centrality of the great towns to debates about cultural issues and social problems in Victorian Britain:

> Manchester, Bradford, Glasgow, Leeds, Liverpool, Birmingham, as well as nineteenth-century London constituted the vanguard of the modern city. Notions of urban life and debates about the meaning of the city which were first developed in these conurbations – the sociology of the city, the politics of the city, the culture of the city, sex and the city – continue to dominate modern approaches to city life some two hundred years on.[4]

The conurbations listed above are reflected in the cinematic city and an in-depth understanding of industrial spaces on film must take into account the physical transformation of the city and its effect on the urban experience, its social identities and relationships.

By the late 1890s, the example set by pioneers like Friese-Greene and Le Prince became common practice and street scenes were filmed throughout the country. Films of provincial England presented themes and subjects in continuity with films of London. For example, royal events were a common

2 Richard Sennett, *The Fall of the Public Man* (London: Faber and Faber, 1986), 2–10.
3 Joyce, 216.
4 Tristram Hunt, *Building Jerusalem: the Rise and Fall of the Victorian City* (London: Orion Books, 2005; first published by Weidenfeld & Nicolson, London, in 2004), 6.

subject both in London and in smaller cities in England, as demonstrated by the film of the visit of Queen Victoria to Sheffield produced by the Riley Brothers in 1897. This local company also produced scenes of everyday street life in Bradford and Leeds. Filmed a few yards away from the location of Le Prince's groundbreaking 1888 film, Riley's *Boar Lane* is a lively street-corner scene showing horse-drawn traffic, trams and pedestrians moving freely in Boar Lane, which was lined with the genteel residences of some of the merchants and traders who contributed to a substantial increase in the city's wealth throughout the century.[5] The cityscape is here composed of a solid built environment containing fluid orderly movement, a combination that consistently characterizes the early cinematic image of the industrial city.

The extensive list of non-extant films from the period 1895–1914 containing references to cities in their titles reveals that the cinematograph reached virtually every British region before the end of the nineteenth century.[6] Most film companies were based in London and filmed both in

5 See Asa Briggs, *Victorian Cities: a Brilliant and Absorbing History of Their Development* (London: Penguin, 1990), 4–10.

6 Birmingham and Leicester were filmed by R.W. Paul in 1896 and 1897 respectively. A market in Nottingham was filmed for the Warwick Trading company in 1900. Worcester was filmed in 1898 by Philipp Wolf for London-based film exhibitor David Devant. Cecil Hepworth filmed a panorama of the city of Bristol from the top of a tram in 1904. The industrial exhibition in Wolverhampton in 1902 was the subject of a film produced by the Biograph Syndicate. The Pageants depicting the histories of Coventry and York were the subject of two films produced by Gaumont in 1907 and Charles Urban in 1909 respectively. Two years later York was filmed by Frank Butcher for Empire Company. Other historic towns were filmed in the early 1910: Oxford by Cricks & Martin and Canterbury by Urban both in 1911, and Guildford in 1913 by J.B. McDowell for the B&C Company. These are the titles of the films listed in the above paragraph: *The Queen at Sheffield* (Riley Bros, 1898), *Bradford Street Scene*, 1898 (Riley Bros, 1898), *Street Scene in Boar Lane* (Riley Bros, 1898), *Birmingham* (Paul's Theatrograph, 1896), *Carting Snow, Leicester* (Paul's Theatrograph, 1897), *Nottingham Goose Fair* (Warwick Trading Co., 1900), *Worcester Street* (Philipp Wolf for David Devant, 1898), *Panorama of the City of Bristol* (Hepwix, 1904), *Wolverhampton Exhibition* (Biograph Syndicate, 1902), *The Coventry Pageant* (Gaumont, 1907), *The York Pageant* (Louis N. Parker for Urban, 1909), *The City of York* (Frank Butcher

the capital and the country: Urban, Paul, and Hepworth, among others, sent their cameramen to film cities all over England. A smaller number of local companies specialized in local material. The already-mentioned Riley Brothers, a company based in Bradford, exclusively filmed in Yorkshire. James Mitchell and Sagar Kenyon filmed rural places, market towns and cities both in Scotland and in many parts of England from Devon to Northumberland. In particular, they filmed in the north of England, with several hundred films made in Lancashire and Yorkshire. As Vanessa Toulmin suggests, the films made by Mitchell and Kenyon contribute to our understanding of the ways in which people in Britain lived during the first decade of the twentieth century, and offer images of leisure, entertainment, sport, work and militarism in the Edwardian age.[7]

Before focusing on Liverpool, we should consider the role of smaller towns and cities on the English coast in the cinematic map of Victorian and Edwardian Britain. These were portrayed as holiday destinations, Brighton and Blackpool, or places of departure, Portsmouth and Southampton. Seaside resorts, as John K. Walton argues, played a central role in the development of tourism as an industry, and as agents of social transformation which made holidays accessible to a larger group of people.[8] At the turn of the twentieth century, seaside resorts were thus filmed as places of amusement and entertainment.[9] Brighton, for example, was filmed by a number of local filmmakers who formed a group usually referred to as the Brighton School.[10] In 1896, Esmé Collings filmed *A Street in Brighton*, *The Promenade*

---

for Empire, 1911), *Historic Oxford* (Cricks & Martin, 1911), *Quaint Kentish Towns: Canterbury* (Urban, 1911), *Guildford* (J.B. McDowell for B&C, 1913).

7 See Toulmin, *Electric Edwardians: the Story of the Mitchell & Kenyon Collection.*

8 See John K. Walton, *The British Seaside: Holidays and Resorts in the Twentieth Century* (Manchester: Manchester University Press, 2000). See also Stephen V. Ward, *Selling Places: the Marketing and Promotion of Towns and Cities, 1850–2000* (London: Taylor and Francis, 1998), 31–32.

9 The Somerset seaside resort Weston-super-Mare, for example, was filmed in 1905: *Bandstand on Weston-super-Mare Pier* (no credits available).

10 See, for example, Frank Gray, *The Hove Pioneers and the Arrival of Cinema* (Brighton: University of Brighton, 1996), and George Sadoul, *Histoire Générale du cinema: 2. Les pionniers du cinéma 1897–1909* (Paris: Denoël, 1948).

at Brighton, Crowds at Brighton and *Boys Scrambling for Pennies under the West Pier*. George Albert Smith made *Brighton Seagoing Electric Car* for his own company in 1897 and, one year later, *Brighton*, a series of scenes on the beach filmed for the Warwick Trading Company.[11] In 1899, Smith also filmed *The Brighton Fire: Arrival of the Brigade*, two shots recording the Brighton Fire Brigade in action. A series of films of Sussex was filmed in 1900 by James Williamson from Brighton, and included the extant scene *Early Fashions on Brighton Pier*, a single shot from a fixed camera position, taken on the pier looking out to the sea and with a group of well-dressed man and women strolling past. The Sussex seaside resort also attracted R.W. Paul, who filmed *Brighton Pier*, *Diving and Bathing in the Sea*, and *Riding on the Beach* in 1897. Paul also filmed scenes of Brighton for his 1898 series *Street Scenes*, which also included views of Edinburgh, Sheffield, Hull, Liverpool and Cardiff. Blackpool, the most popular centre of tourism for the inhabitants of northern mill towns, was the subject of several films. Hepworth's *Blackpool: Panorama of the Sea* (1899) and *Panorama of Blackpool* (1904) were both filmed from the top of a moving tram. In Hepworth's *Blackpool: the Parade* (1899) the Blackpool Tower, which was opened to the public in 1894, is visible in the background, with a large crowd occupying the foreground of the shot. A famous achievement of Victorian seaside architecture, the Tower was also filmed by Mitchell and Kenyon.[12] *Blackpool Aviation Week*, which took place between the 18th and 23rd of October 1909, was the subject of films produced by Walturdaw, Paragon and Kineto. These films appeared when seaside tourism was established as an industry appealing to both the working class and the middle class. In the nineteenth-century people living in the largest cities of Britain experienced an increasing freedom of movement in the city, across urban and regional boundaries; films of Blackpool record the presence of visitors experiencing leisure allowed by the freedom to move.

11 The series included: "Children on the sands", "The ungallant lover", "The esplanade", "Launch of the Skylark" and "The lower esplanade."

12 Toulmin, *Electric Edwardians: the Story of the Mitchell & Kenyon Collection*, 93.

Coastal towns with a large marine and naval role were also the subject of several films. Here, the theme of movement was implied as the act of crossing national boundaries sailing to distant destinations – including colonies, dominions and the battlefields of the Boer War. In 1896 Esmé Collings filmed *A Street in Portsmouth*, *Portsmouth: the Ferry* and *Soldiers in Portsmouth*, a scene showing soldier departing for South Africa. Two years later, George A. Smith also filmed the ferry, and the London-based company Chard's Vitagraph made *Portsmouth Fire Brigade*. Plymouth was first filmed in 1898 by Cecil Hepworth for the Warwick Trading Company (*Panorama of Plymouth Harbour*) and ten years later by Charles Urban in the series *Salisbury to Plymouth*, which also included views of Axminster, Exeter, Okehampton and Dartmouth. A port city like Southampton, which by 1914 was the fifth most important British port in terms of shipping tonnage, was often filmed in connection with the departure and return of the troops to and from the war in South Africa. Films of the City of London Imperial Volunteers embarking at Southampton were produced by Paul's Theatrograph, Biograph Syndicate and the Warwick Trading Company. *The Lord Mayor at Southampton*, *City of London Imperial Volunteers Embarking* and *CIVs Embarking at Southampton* were all filmed in 1900, and record the progress of CIVs from London to Southampton. Hepworth filmed for the Warwick Trading Company *Kitchener's Arrival at Southampton* (1900), which has survived in an incomplete print and shows a city street decorated with banners reading "Upholders of the Empire" and "Welcome Home", with the railway station in the background. The street is lined with troops and crowds and a procession of carriages and cavalry passes through. Lord Roberts, Lord Kitchener, General John French and General Ian Hamilton are also part of the procession leaving from the Harley Institute (now the University of Southampton). The cinematic image of Southampton in such films was influenced by military events in distant parts of the empire, including South Africa; it was also defined by scenes of everyday life on the streets of the city such as those offered by *Tram Journey Through Southampton*, a film produced by the British Mutoscope and Biograph Company in 1900. The camera is here fixed on a tram heading through an archway, a section of the remains of the medieval stone wall, and travelling south on Southampton High Street, which is lined with shops and crowded

with men, women and children. Two other trams and two carriages can be seen passing on the adjacent line, which is separated by a row of lampposts. The cinematic image of Southampton is thus made of everyday activities and imperial iconography in the form of military events. Films of Southampton and Portsmouth conveyed meanings associated with the military venture in South Africa and thus with the expansion and preservation of Britain's colonies. These themes and a significant focus on mercantile dynamism emerge on a greater scale from films of Liverpool and its port.

Cinematic Liverpool: Gateway of the British Empire

The docks in east London were filmed on very few occasions. R.W. Paul filmed *Unloading a Cargo at London Docks* in 1898, and Charles Urban made *Swimming and High Diving at East India Docks* in 1905. There is hardly any other trace of films of the London docklands in the early days of cinematography – except for the glimpses of the docks seen through the newly-built Tower Bridge in films made by Promio and other filmmakers. The Port of Liverpool, on the other hand, was filmed more often and its cinematic image gives us the opportunity to investigate the imperial theme in relation to the movement of people and things from Britain to the colonies and vice-versa. In the early 1870s, Robert Dudley painted *Canada Timber Docks, Liverpool, towards close of day*. This painting is a charming view of the hustle and bustle of the port. Canada Dock opened in 1858, when Canada was Britain's major source of timber, and was designed and built by civil engineer Jesse Hartley. Dudley portrayed a number of sailing ships lined-up on the dock sides crowded with workers unloading wood stacked on the quaysides and carted away by horses. Imperial iconography, as it emerges from the built environment and urban events such as parades and processions, defined early cinematic Liverpool as the gateway of the British Empire – this perspective was reflected in film.

In 1901 Liverpool had a population of almost seven-hundred-thousand people. Throughout the nineteenth century the city expanded in all directions and new docks and residential districts appeared. This process of urban expansion is well-described by Francis E. Hyde:

> In the period 1850 to 1900, terraced houses, hideous in the yellow brick of the Welsh building contractor, had topped the sandstone ridge and were engulfing Everton and Wavertree. Long ribbons were stretching out across the fields and heathlands to West Derby, Childwall, Gateacre and towards the pleasant 1820 resort of Waterloo; and as the demand for ships increased so too did the demand for docks. Northwards and southwards from the original eighteenth-century dock system new wharves and quays, new wet-docks and dry-docks appeared. Behind them rows of mean streets, housing dock workers, were already engulfing the once pleasant villages of Bootle and Seaforth to the north and St Michael's hamlet, Toxteth, and Dingle to the south. Liverpool had extended her influence to the far ends of the earth, she had amassed vast wealth, but she had become ugly in the process.[13]

The expansion of the city and its population was a result of the increasing importance of Liverpool in the British sea trade. The suburbs, as Hyde explains, became uglier throughout the century, while the monumental city centre acquired handsome buildings which reflected the importance of the city, wider well-lit streets and a new sewage system. At the turn of the twentieth century, the port of Liverpool accounted for 31 per cent of the United Kingdom's visible imports and exports, and was regarded as one of the main ports of the empire.[14] As Sarah Palmer suggests, the port of Liverpool was export-oriented and drew on the manufacturing and industrial strengths of its hinterland. Its position on an exposed near-coastal site, with excellent canals, railways and roads, the extensive dock and quay facilities as well as the wide range of waterfront ancillary services and maritime-related industries, contributed to making Liverpool one of

13 Francis E. Hyde, *Liverpool and the Mersey: the Development of a Port 1700–1970* (Newton Abbot: David & Charles, 1971), 69–70.

14 See Tony Lane, *Liverpool: Gateway of Empire* (London: Lawrence & Wishart, 1987), 36. See also Gordon Jackson, "Sea trade", in *Atlas of Industrializing Britain, 1780–1914*, 94–105.

the great ports of the empire.[15] All kind of people and goods came and went to and from the colonies, making the port of Liverpool one of the most diverse places in Britain. Written accounts from the late nineteenth century are rich in fascinating description of Liverpool and the docks on the River Mersey, and the frenetic activities taking place on the wharves and quays. In 1881, the Reverend Fred Bell described the sight of the port as a "long vista of vessels alongside the quay, lashed together with planks, reaching far ahead." The atmosphere of the port emerges from Bell's description as he continued:

> Multitudes of goods are shipped – all kinds of hardware, railway supplies, iron in all shapes, of all kinds and sizes, sheet, wire, bar, spring, etc.; bales, boxes, casks, wines, spirits, ales, for India, Madagascar, Asia, Persia, the Continent and America... Thousands of men are here measuring packages, invoicing goods, shipping merchandise; the tramp of horses, songs of stevedores, and shouts of sailors make a very Babel of industry.[16]

Colour, variety and diversity defines this description of the Port of Liverpool. The port was a pivotal point on the map of the empire and also a vital element in the economic and social life of the city. In 1891, 27 per cent of the male population of Liverpool over the age of ten worked in transport-related industries. Over 30 per cent of total imports were cotton, while cotton yarns and other textiles amounted to 50 per cent of the total imports by value. Between 1860 and 1900 almost five million British people migrated to America from Liverpool, although passenger traffic began to decline and was largely moved to Southampton early in the twentieth century.[17] Beyond the financial centrality of the port, the energetic activity on the docks naturally attracted the attention of painters

15 Sarah Palmer, "Ports", *The Cambridge Urban History of Britain*, Vol. II and III, ed. Peter Clark (Cambridge: Cambridge University Press, 2000), 140.

16 Fred Bell, *Midnight Scenes in the Slums of New York* (New York: John Howitt and Son: 1881), 230–231.

17 Joan Smith, "Class, skill and sectarianism in Glasgow and Liverpool, 1880–1914", in *Class, Power and Social Structure in British Nineteenth-Century Towns*, ed. Robert J. Morris (Leicester: Leicester University Press, 1986), 168–169.

like Robert Dudley and writers like Fred Bell. A few years before Bell wrote his account, Revd Francis Kilvert thus described the Mersey:

> The Mersey was gay and almost crowded with vessels of all sorts moving up and down the river, ships, barques, brigs, brigantines, schooners, cutters, colliers, tugs, steamboats, lighters, "flats", everything from the huge emigrant liner steamship with four masts to the tiny sailing and rowing boats. From the river one sees to advantage the miles of dock which line the Mersey side, and the forests of masts which crowd the quays.[18]

These descriptions highlight both the centrality of Liverpool on the map of the empire and the centrality of the port to the city. The historian Tony Lane has drawn attention to the importance of these connections. Unlike a cotton town such as Manchester and its regularly employed factory workers, Liverpool required a large and casually employed population to work in the docks. Liverpudlians, Lane explains, were employed to sail the ships, repair them, load, store, shift and process their cargoes. This had a great impact on the economy of Liverpool, and seasons, weather and tide affected employment in the city: when the ships were in there was employment and when the docks were slack workers were laid off. The city, Lane suggests, was also given colour and cosmopolitanism by a transient population of seafarers.[19] This variety is reflected in visual and written descriptions of the city and can be seen in early moving images of Liverpool.

The great scale of the Port of Liverpool attracted pioneer filmmakers as did the monumental spaces of London, and as in the capital of the empire the hand of the cameramen created a selective image of the city through omission and emphasis. As I have argued above, films signify by their exclusions just as by their inclusions. The commercial centre of Liverpool, the port, and the monumental city centre were the subject of several early actuality films. On the other hand, the slums and the poor were structurally absent from representation like the neglected urban spaces in the East End of London. As Krista Cowman has argued, the proximity of

18 Francis Kilvert, in *Kilvert's Diary, 1870–1879*, ed. William Plomer (Harmondsworth, London: Penguin, 1982), 193.

19 Lane, 18.

the slums to the mercantile centre and the fine cluster of buildings of the city centre – which included a new Town Hall, the neo-Grecian St George's Hall and the Stock Exchange – revealed the differences between the rich and the poor inhabitants of the city.[20] In 1901, a journalist from *The New Penny Magazine* also noted that in Liverpool "sharper contrasts between the grandiose and the squalid than are generally to be observed."[21] No trace of such contrast can be found in early films, which focused on the docks and the monumental city centre, thus creating an image of a monumental city with trade connections to the empire.

The Port of Liverpool was first filmed in 1897 by Alexandre Promio. *Panorama of Liverpool Docks Taken from the Electric Railway* was filmed from a train on the Liverpool Overhead Railway, the world's first electrically-operated railway which opened in 1893 and ran along the docks from Dingle, on the south end of Park Road, to Seaforth Sands.[22] Promio recorded busy activity on the docks and the warehouses, and in *The Harbour* he filmed the Mersey from Liverpool to Birkenhead, a town on the Wirral Peninsula, along the west bank of the River Mersey and opposite Liverpool: the camera pans from left to right and a sailing boat, a tug and a steamship with four masts and one funnel can be seen crossing the screen from right to left. A third film, *Entry to Clarence Dock*, was filmed from a boat nearing the self-contained steamship dock facility named after William IV: the camera pans left to right and records a working day on the dock where goods are shipped, revealing a busy workplace. Promio was succeeded by his British colleagues who later filmed the same places in the port of Liverpool. In 1897, Chard's Vitagraph produced *Woodside Ferry*, which was filmed in

20 Krista Cowman, "The battle of the boulevards: class, gender and the purpose of public space in later Victorian Liverpool", in *Identities in Space: Contested Terrains in the Western City since 1850*, eds Simon Gunn and Robert J. Morris (Aldershot, Burlington, Singapore and Sydney: Ashgate, 2001), 152.

21 Anon. (1901), "A commercial wonderland: Liverpool and its docks", *The New Penny Magazine*, no. 109, vol. IX, 177–184.

22 For an in-depth architectural study of Promio's films, see: Richard Koeck, "Liverpool in Film: J.A.L. Promio's Cinematic Urban Space", *Early Popular Visual Culture* 7:1 (2009), 63–81.

Woodside, Birkenhead, almost opposite Liverpool Pier Head. This film represented the steam-powered vessels, known as "twin screw" steamers, which connected Woodside to Liverpool.[23] In the same year, Paul's Theatrograph produced *Ferries Arriving at Landing Stage*, a film recording the arrival in Liverpool of these steam-powered ferries. The wreck of Admiral Nelson's old flagship, HMS Foudroyant, was filmed in Vitagraph's *Woodside Ferry*. In 1899, the Warwick Trading Company produced *Liverpool from River Mersey* and a series called *R.M.S. Oceanic*, which recorded the world's largest ship entering River Mersey on her maiden voyage from Belfast and included "Arrival of the Oceanic at Liverpool" and "Panorama of Liverpool." One year later, Warwick used – as Promio had – the elevated position of the Liverpool Overhead Railway and produced two films: *Glimpses on the Liverpool Overhead Railway* and *The Electric Trains of the Liverpool Overhead Railway*. Well into the first decade of the twentieth century, the docks of Liverpool were still an attractive subject for filmmakers. Joe Rosenthal, who had left the Warwick Trading Company in 1908 and set up his own company, Rosie Films, based in Croydon filmed a visit to the Port of Liverpool in 1910. During that year, the same subject was filmed by Walter Tyler in *Liverpool: the Gate of the West*. These films are now lost but similar scenes can today be seen in the recently found films of Mitchell and Kenyon.

Until the discovery of the Mitchell and Kenyon Collection, the films made by Alexandre Promio were the only moving images of the Liverpool docks at the turn of the twentieth century known to have survived. In 1901, Ralph Pringle, the manager of the Liverpool-based *North American Photo Company* which managed film distribution from the Midlands to Scotland and down to the South West, commissioned Mitchell and Kenyon to produce a series of films of local interest to be shown as part of a larger programme of entertainment exhibited to the workers and locals.[24] Mitchell and Kenyon mainly filmed the docks and great events in the heart of the city, but also a number of football matches played by local teams Liverpool and Everton as well as a few local events like church parades.[25]

23 See Hyde, 52.

24 Toulmin, *Electric Edwardians: the Story of the Mitchell & Kenyon Collection*, 57–84.

25 Titles: *Liverpool v. Small Heath*, *Everton v. Newcastle United* and *Everton v. Liverpool*, *Liverpool Church Parade and Inspection*. Sports were a relatively frequent subject in

The docks were a recurring subject of representation in the films made by Mitchell and Kenyon for Pringle. *Employees Leaving Alexandra Dock* was filmed by a main thoroughfare with a never-ending flow of carts and male workers at the entrance to Alexandra Dock, which was mainly involved in the grain and cotton trade and also provided refrigeration facilities for imported frozen meat. The activities on the docks are not at all clear from the limited view provided here by the cameraman. However, despite the emphasis placed by Mitchell and Kenyon on capturing as many human faces as possible, a sense of the large scale of the docks operation is suggested by the long shot of a large number of workers emerging from the entrance opposite the camera position, and by the presence of heavy-duty carts carrying cotton bales.

The docks feature in two other films by Mitchell and Kenyon. *Liverpool as seen from the Front of an Electric Car from the Pierhead to the Circus* is a panoramic view of the docks and the city, and includes a view of the roofs, a church and crowded streets. The electric car follows a route from the Pierhead – filmed here before the construction of its three iconic buildings, the Royal Liver Building, the Cunard Building and the Port of Liverpool Building – through the crowded and busy city centre, and down to the Circus on West Derby Road. This film shows how vast the port was and offers a view of the scale of the mercantile activity in the area. *A Trip to North Wales on the St Elvies* represents the St Elvies, a paddle steamer operated by the Liverpool and North Wales Steamship Company, and it shows passengers on the ship and in the boarding area. It also includes a view of the Liverpool docks filmed from the ship. Mitchell and Kenyon recorded images of an urban space which was becoming a powerful symbol of imperial ambition and urban modernity thanks to the combined forces of municipal enterprise and mercantile profit.[26]

---

early topical cinema and the company named Archer & Sons filmed *Liverpool Police Sports* as early as in 1898. A later film on Liverpool police was produced in 1901 by the Warwick Trading Company and it is now lost: *Review of the Liverpool Police*. On Mitchell and Kenyon sport films, see: Toulmin, *Electric Edwardians: the Story of the Mitchell & Kenyon Collection*, 126–153.

26 On the three buildings of the Liverpool waterfront, see: Peter De Figueiredo, "The Buildings of the Liverpool Waterfront", *Architectural History*, vol. 46 (2003), 229–254.

Ferries and ships on the Mersey can be seen in other films by Mitchell and Kenyon for Ralph Pringle. *Prince of Siam aboard the New Ferry* shows the smartly-dressed *Crown Prince Vajiravudh*, his aides and British dignitaries on the Wirral Peninsula, on the deck of the Birkenhead Ferry on their way to Port Sunlight. *Cunard Mail Steamer Lucania Leaving for America* consists of a series of scenes showing wealthy passengers and staff disembarking from a journey on the *RMS Lucania*, which could carry over 2,000 passengers en route from Liverpool to New York. The Captain and other senior officers can be seen walking along the deck towards the camera. *SS Saxonia in Liverpool* represents the Cunard steam ship Saxonia returning from Boston through Queenstown (now Cobh) in Ireland. The Saxonia had just become a Royal Mail Ship, and dock workers can be seen unloading sacks of mail. *SS Skirmisher in Liverpool* also shows passengers boarding a ship, deck-hands cleaning the decks with a hose, and the Captain walking on the deck.

These films show workers and other staff engaged in their quotidian activities at the docks. They also focus on the ships and ferries landing and departing from the port and thus suggest, by defining the Port of Liverpool as the centre of a centripetal and centrifugal force, the theme of movement of people and things to and from other distant territories of the British Empire and other parts of the world. Early films thus reflect the modernity of maritime transport in Liverpool in relation to technological shipping developments such as the employment of steam turbines and multiple expansion engines. On those ships, as well as those departing from London or Southampton, British filmmakers left Britain to explore and film distant cities of the empire, and on those ships they returned to Britain with films of such places to be distributed and exhibited in British cities, a subject I will return to in the following chapters.

Pioneer filmmakers also focused on the centre of Liverpool, which emerges as a solid built environment defined by that increasing ease of movement which characterized late Victorian and Edwardian cities. Alexandre Promio filmed *Church Street* on his visit to the city in 1897. Today Liverpool's main shopping area, Church Street lies between Bold Street to the East and Lord Street to the West and was filmed diagonally from the intersection with Bold Street. *Church Street* opens with the passage of a tram going to West Derby and closes with a second tram going towards the Pier Head. A shot of a number of men crossing the road in top hats and long coats shows that Church Street was a wide and busy street where many shops with large awnings were located. Four years later, Mitchell and Kenyon filmed *Liverpool Street Scenes* in Holy Corner, at the intersection of Church Street, Whitechapel, Paradise Street, and Lord Street. The scene includes views of the old Compton Hotel and the Bunney's department store.[27] In this film this the inner districts of Liverpool emerges as wealthy places where trade was mostly conducted on busy thoroughfares with the shop fronts filled with goods for sale. These street scenes were taken in the commercial district of Liverpool, which had been redeveloped in the 1850s and was characterized by wide streets with modern lighting, sewage and water supplies.[28] Films of Liverpool thus display a city where people, vehicles and goods could circulate freely: this ease of movement was exercised by the inhabitants of the city and facilitated by improvements in the built environment and public transport.

Lime Street, a few yards away from the Holy Corner, was filmed by Promio in 1897. His *Lime Street* is another crowded street scene with horse-drawn trams passing by, and several men and women crossing. The Liverpool Corporation ran the horse-drawn tram line and opened its first stretch of electrified tram lines in 1898. By 1902 the whole system was converted and extended into the newer suburbs.[29] Promio had thus recorded horse-drawn transport in Liverpool in its very last days. Just like the modern steamers, electric trams came to represent the modernity of new development in urban transport made possible by the wealth accumulated by the city thanks to the prosperity of its trade. Later films, such as those made by Mitchell and Kenyon, record this development. More prominently, *Lime Street* is dominated by the monumental presence of the neoclassical building of St George's Hall, home to concert halls and law courts, and St John's Church (which closed in that year and was later demolished). St George's Hall, which can be seen in the background of this shot, is one of the finest

27 See Toulmin, *Electric Edwardians: the Story of the Mitchell & Kenyon Collection*, 163.

28 See Cowman, 150–152.

29 See Theodore C. Barker, "Urban Transport", in *Transport in Victorian Britain*, Michael J. Freeman and Derek H. Aldcroft (Manchester and New York: Manchester University Press, 1988), 157.

neo-Grecian buildings in the world and, as Tristram Hunt suggests, it can be seen as a testimony to Liverpudlian self-confidence.[30] Bearing in mind Henri Lefebvre's argument that the image of urban membership is to be found in the symbolic underpinning of a society created by monuments and landmarks, moving images of St George's Hall suggest an undeniable attraction felt by middle-class filmmakers for monumental sites in the city.[31] St George's Hall gave Liverpudlians an urban symbol to identify their city with: films of this building made Lime Street immediately recognizable on the screen.

Just like the mercantile dynamism of the docks and the commercial core in the Holy Corner, landmarks such as St George's Hall projected an image of the city's self-confidence. On the other hand, the built environment of the city centre acquired an imperial dimension during military and royal parades and processions taking place on the streets. Festivals and rites can be seen in terms of symbolic systems engaged in a continuous creation of political and social meaning, and films contributed to sharing information on the visual power of public events. A small number of films made by Mitchell and Kenyon for Ralph Pringle depict imperial, royal and military events which defined the ways in which politics extended beyond the ruling elite and assumed new shape and significance on the city streets. For example, *Princess Louise at Liverpool* shows a kilted regimental bagpipe band performing, as the Lord Mayor and Princess Louise enter a carriage. Events like royal visits and celebrations like Trafalgar Day and St George's Day or the Liverpool Pageant, which was filmed in 1907 by Gaumont, had a strong imperial and civic significance. Just like Queen Victoria's Diamond Jubilee procession and other events filmed in London, they contributed to making the city more fittingly imperial by displaying military heroes and colonial subjects, by decorating the city buildings with the Union Flag and other national symbols.[32]

30 Hunt, 235–236.

31 Lefebvre, *The Production of Space*, 220–221.

32 *The Liverpool Pageant*, Gaumont (1907).

The Second Boer War was a theme implicit in films such as those portraying the City Imperial Volunteers in London and Southampton. The war in South Africa also affected the cinematic image of Liverpool as troops returning to Ireland from South Africa also landed in Liverpool. Though never as violent as those of Belfast or Glasgow, Liverpool had its religious controversies and was a centre of sectarian bitterness. A permanent change in the social structure of the city, with Great Homer Street dividing the Catholic and Protestant districts, resulted from Irish immigration to the city following the famine of 1845–1849 and emerged as an important factor in engendering sectarian discontent. At the turn of the century, the Boer War engendered nationalist fervour which worked against anti-Irish feeling.[33] On November 24th, the *Liverpool Mercury* reported the decision taken by the Lord Mayor to allow the use of St George's Hall to entertain the 5th Irish Volunteers, and the invitation by Lieutenant-Colonel B.J. Robinson to the people of Liverpool to "parade in a welcome to the home-coming volunteers" on the following Wednesday.[34] The only film of troops returning to Liverpool is *Return of Liverpool Yeomanry from the Front*, which was produced by the Warwick Trading Company in 1901, and recorded the welcome given by the people of Liverpool to units of the British Territorial Army. Scenes like these were increasingly familiar in British cities, and a popular subject of non-fiction films.

The Boer War also defined the representation of central Liverpool in two films made by Mitchell and Kenyon in the same year, which present the visit of the Anglo-Irish Lord Roberts in a kind of hero-worship iconography. Lord Roberts, the commander of the British troops fighting in the Boer War, appeared in *Visit of Earl Roberts and Viscount Kitchener to Receive the Freedom of the City* and *Lord Roberts Presenting Medals to Boer War Volunteers in Liverpool*. The son of an Irish couple and born in India in 1832, Lord Roberts was a quintessentially imperial figure and his

33 See Eric Midwinter, *Old Liverpool* (Newton Abbot: David & Charles, 1971), 170–172; Frank Neal, *Sectarian Violence: the Liverpool Experience, 1819–1914: an Aspect of Anglo-Irish History* (Manchester: Manchester University Press, 1988), 196; Smith, "Class, skill and sectarianism in Glasgow and Liverpool, 1880–1914", 198.

34 *Liverpool Mercury* (Liverpool, England), Monday, November 24, 1900, p. 1.

presence haunted the cinematic city at the turn of the century. He was filmed in London on his return from South Africa and during the Diamond Jubilee, in Liverpool, Manchester, Aberdeen and Cape Town. In *Visit of Earl Roberts and Viscount Kitchener to Receive the Freedom of the City*, in which a great civic parade is watched by a large number of spectators, Roberts and Viscount Horatio Kitchener attracted a large crowd onto the streets of Liverpool and, despite growing anti-war feeling, gave many Liverpudlians a reason to celebrate national achievements.[35]

As is revealed by the sight of the entrance to the Empire Theatre, *Lord Roberts Presenting Medals to Boer War Volunteers in Liverpool* was filmed on Lime Street, near St George's Hall. It features a large crowd on the pavements and on the trams awaiting the arrival of Lord Roberts, while a band, troops and veterans with medals on their uniforms parade on a rainy and windy day. Both St George's Hall and the equestrian statue of Prince Albert, erected in 1866, are visible in the shot. As Vanessa Toulmin suggests, films like *Lord Roberts Presenting Medals to Boer War Volunteers in Liverpool* were all marked by crowds and civic parties in the heart of the city and reflect the city's fondness for parades and festivities, especially of a military and patriotic kind.[36]

Mitchell and Kenyon also filmed the celebration of St George's Day and Trafalgar Day in the central district of Liverpool. Lime Street features again in *St George's Day Procession in Liverpool*, which consists of a number of shots of the crowded street and its monumental palaces. Bands of young musicians march down a street dominated by the Union Flag and the St George's Cross. In 1897 the celebrations for the victory of Vice-Admiral Horatio Nelson's British fleet over the French and Spanish fleets at the Battle of Trafalgar in 1805 had been filmed in London by Mutoscope and Biograph, the Warwick Trading Company and W. Watson & Sons (see Chapter 2). Trafalgar Day was a very popular event until the outbreak of the First World War and was celebrated throughout the country. In Liverpool, the celebrations were normally organized by the Navy League,

35 See Toulmin, *Electric Edwardians: the Story of the Mitchell & Kenyon Collection*, 171–173, 258–260.

36 See Toulmin, *Electric Edwardians: the Story of the Mitchell & Kenyon Collection*, 169–170.

and in 1901 were filmed by Mitchell and Kenyon for the *North American Photo Company*. In 1899, the annual event was described thus on the pages of the *Liverpool Mercury*:

> The patriotism of the city reached the highest point on Saturday. There was demonstratively celebrated the anniversary of the Battle of Trafalgar and its co-incident, the death of heroic Nelson. Noon was the hour appointed for the ceremony of the placing by right Hon. the Lord Mayor of a commemorative wreath upon the famous monument of the great naval commander, which is so conspicuous a feature of the spacious quadrangle which is known colloquially as the Exchange Flags, but long prior to this a densely compact mass of people had taken up a position there. The upper portion of the portico which occupies three sides of the Exchange, the latter an imposing and elegant structure, was occupied by a throng of ladies and gentlemen, while the window of that and other buildings in the immediate neighbourhood were similarly occupied. The popular emblem of British supremacy, the Union Flag, was everywhere stirred into graceful movement by a gentle breeze, while a semi-wintry sun gave a golden glow to the picturesque scene of animation.[37]

This article points out the character of the celebrations for the anniversary of the Battle of Trafalgar, and refers to the main features of this and other events: large masses, monumental spaces, formal rites and icons such as the Union Flag. Two years later these characteristics emerged from the film shot by Mitchell and Kenyon on the very same location and including a view of the Exchange Flags with the portico and the Doric columns of the classical three-storey branch building of the Bank of England. *Trafalgar Day in Liverpool* was shot in Castle Street at the base of Nelson's Monument decorated with Union Flags, and shows dignitaries, officers, crew and cadets marching on the street. The events filmed by Mitchell and Kenyon were characterized by the presence of large crowds and orderly movement of men in uniform in the monumental city centre. The popularity of these celebrations and rites, both with the urban population and with cinema audiences, reveals the importance of imperial, national and patriotic themes and feeling in the lives of the British people in the age of high imperialism.

37 "Trafalgar Day in Liverpool: Celebration of the Anniversary – A Patriotic Demonstration", *Liverpool Mercury* (Liverpool, England), Monday, October 23, 1899, p. 1.

On a cinematic map of the British Empire, Liverpool appears a space of connection and the city is defined by commercial, ceremonial and military dynamism. Most films of the city centre and its monuments reveal the civic pride and confidence of the city authorities, and the festivals and rites taking place on the streets of the central district further emphasize the important role of nationalism and imperialism in the modern experience of the city. Developments in transport, expressed by films of the electric overhead railway and the electric trams which replaced the horse-drawn trams filmed by Promio in 1896, reveal an urban space where greater speed and ease of movement were possible. Liverpool can thus be rightly seen as a truly modern city defined in great part by its mercantile and commercial power. Nothing seemed to be able to diminish the might of the Port of Liverpool and, as Francis Hyde suggests, even the opening of the Manchester Ship Canal in 1894, which could have made Manchester emerge as a rival port to Liverpool, resulted in a non-existent rivalry.[38] As I shall demonstrate in the next section, films of Manchester equally expressed the economic aims of imperialism. However, Manchester was the archetypal mill town, Liverpool one of the great ports of the empire. This difference was also reflected in films.

Manchester: Films of industry and civic pride

In 1896 Manchester had its first public screenings of moving pictures, which took place at the St James Theatre in Oxford Street, and the old YMCA building and the Lesser Free Trade Hall, both on Peter Street.[39] Film pioneers began filming the city shortly after the first screenings took place. Both local and national film companies filmed Manchester's city centre and, as in London or Liverpool, avoided its poorest working-class

38 Hyde, 134–138.

39 Alan J. Kidd, *Manchester* (Keele: Ryburn, 1996), 130.

districts, which had famously been represented by Samuel Coulthurst in the early 1890s in a series of photographs of buildings, streets and people.[40] In 1898, the Warwick Trading Company produced a film containing busy street scenes in Corporation Street and Market Street and three years later its cameramen filmed the famous horse-race the Manchester Cup.[41] Unfortunately, these films are now lost and the only recorded moving pictures of Edwardian Manchester come from the Mitchell and Kenyon collection. In their films, employees of local factories, engineering works, mills and co-operatives – and potential audiences – were filmed during breaks or at the end of their working day.[42] Early in the nineteenth century the urban space of Manchester had been one of the true epitomes of industrial modernity, and in 1900 the city was still a centre of utmost importance on the map of the British Empire.

In 1852, the construction industry journal *The Builder* reported the words of Sir James Fitzjames Stephen:

> Manchester is a name of deep and even of awful significance for here is the metropolis of that Titanic industry on the continued success of which England has deliberately pledged her station and her authority among the nations of the world.[43]

40 Chris Makepeace (ed.), *Manchester and Salford in the 1890s: A Collection of Photographs by Samuel Coulthurst*, Hendon Mill: Hendon Publishing, 1986.

41 Titles: *Manchester* and *The Manchester Cup*.

42 Tiles: *Employees Leaving Crossley's works, Openshaw, Manchester*, *Workpeople Leaving Ryland's Mill, Gorton, Manchester*, *Emplyees of Galloway Ltd Boiler Works, Hyde Road, Manchester*, *Employees of Co-operative Wholesale Society Printing Works, Lonsight, Manchester*, *Workmen Leaving Brooks and Doxey, West Gordon Works, West Gordon Works, Manchester*, *Workers at Berry's Blacking Works, Manchester*, *Workers at Brooks and Doxey, West Gorton Works, Manchester*, *Workmen Leaving Peacock's Works at Meal Time, Gorton, Manchester*, *Employees of J. & E. Waters, Talbot Mills, Ellesmere St, Manchester* and *Co-operative Wholesale Clothing Factory in Manchester*. See also Tom Gunning, "Pictures of Crowd Splendour: the Mitchell and Kenyon Factory Gate Films", in Toulmin, Popple and Russell *The Lost World of Mitchell and Kenyon*, 49–58, and Toulmin, *Electric Edwardians: the Story of the Mitchell & Kenyon Collection*, 210–211.

43 Sir James Stephen, cited in *The Builder*, vol. 10 (1852), 580.

On the other hand, Manchester was a socially divided urban space. In *The Condition of the Working Class in England*, Friedrich Engels had given an account of mid-nineteenth-century Manchester that was central to the intellectual and imaginative vision of his work as a whole. Engels investigated the industrial might of Manchester and drew attention to the poorest slums of the city. He thus describes a stroll from Old Church to Long Millgate:

> Here one is in an almost undisguised working men's quarter, for even the shops and the beerhouses hardly take the trouble to exhibit a trifling degree of cleanliness. But all this is nothing in comparison with the courts and lanes which lie behind, to which access can be gained only through covered passages, in which no two human beings can pass at the same time. Of the irregular cramming together of dwellings in ways which defy all rational plan, of the tangle in which they are crowded literally one upon the other, it is impossible to convey an idea.[44]

Engels reported the irrational look of the built environment in the slums and emphasized the difficulty of movement in their narrow lanes. In contrast, the central districts of the city saw an increasing ease of movement brought about by the widening of their streets, the development of modern form of transport, and new lighting and sewage systems. At the end of a century in which Manchester took a leading role in the British economy, the city was still characterized by a tension between industrial and commercial dynamism and social inequality. The words of Engels and Stephen focused on Manchester's social problems and the city's industrial and commercial might respectively. As we shall see, films from the period 1895–1914 exclusively explore the latter.

Between 1851 and 1901 the population of Manchester doubled to 645,000 and, as Dennis Richard suggests, it had a rather complex social structure, with merchants, tradesmen and professional people to mediate between mill owners and factory workers.[45] The city swallowed up the surrounding townships to cover by the turn of the century an area of over 300 square miles. On the other hand, the central districts of Market Street,

44 Engels, 87–88.

45 Richard Dennis, *English Industrial Cities of the Nineteenth Century: A Social Geography* (London and Cambridge: Cambridge University Press, 1984), 17.

Deansgate and London Road saw a decrease of their population from over 92,000 in 1851 to 30,000 in 1901. During those years the commercial core of the city expanded as shops, warehouses and banks lined former residential streets and, on the other hand, railway and street improvements contributed to making the city centre a business and market zone. Alan Kidd has drawn attention to the result of the constant street improvements on the new topography of the city:

> Corporation Street had cut a swathe through the rambling byways of the old town in 1845. Portland Street was extended to meet Oxford Street in the 1850s. Albert Square was created in the 1860s as a prelude to the building of the Town Hall in the 1870s. Whitworth Street was laid out and lined with its distinctive terracotta warehouses in the 1890s. By now the central business district incorporated several discrete zones. The warehouse mansions of the cotton merchants around Portland Street contrasted with the more workmanlike premises of the wholesale provision traders at the Withy Grove/Fennell Street end of town where the impressive headquarters of the Cooperative Wholesale Society dominated the northern stretch of Corporation Street. The upper reaches of King Street and Spring Gardens had taken on a sober and serious character appropriate to the city's banking and insurance quarter.[46]

The number of firms occupying office space in the central districts of Manchester increased by 41 per cent between 1871 and 1914 and the number of attorneys and solicitors increased by 60 per cent between 1861 and 1881, with a threefold increase in the number of accountants and commission agents over the same period.[47] These changes in the social structure of central Manchester resulted in an equally remarkable increase of the number of dining rooms and cafes in the city centre. Early filmmakers recorded these changes by presenting a central district occupied by a great number of businesses, shops and cafes, and a rapidly moving crowd accessing the city centre and its services.

By the middle of the nineteenth century, as Simon Gunn has suggested, moral imperatives drove great investments in the building of disciplinary institutions such as courts, prisons, workhouses and asylums, and a drive to

46 Kidd, 138–139.
47 See Kidd, 104–106.

widen and cleanse the city's streets. Local authorities were also determined to clear population from the central core, and at least 30,000 people were displaced in the 1860s in the vicinity of London Road, Market Street and Deansgate. Cleansing the streets was also clearing them of the so-called "residual" population associated with slums. From the 1820s, the city's business elite had in fact withdrawn from the central districts of the city to the suburbs and this process resulted in an early example of social segregation by residence and an empty and functional heart in the middle of the city focused on the exigencies of distribution, consumption, and exchange. On the other hand, Gunn suggests, slums populated by the poor, including the Irish and Jewish communities, impinged on central streets such as Deansgate and Shudehill.[48]

Manchester's central district, characterized by order, wealth and fluidity, emerges vividly from Mitchell and Kenyon's *Manchester Street Scenes*, which was filmed in 1901 at the busy traffic intersection of Corporation Street and Cross Street. In this film we can see a policeman directing horse-drawn traffic – a few months before the introduction of the first electric trams in Manchester – that includes different types of carriages and trams going to Cheetham Hill, Longsight and Hightown. Large groups of men, women and children can be seen by the junction on both sides of the street lined with shops. The scene is characterized by rapid movement: the traffic warden on the left side of the frame suggests the presence of order within the mobility of city life. The trams and carts move in orderly lines in two directions, and the people walk at the side of the street. On the other hand, no pavements are visible and several children and a few adults rapidly cross the street at their own risk. Overall, the frenetic dynamism of this scene characterizes the city as one of the foremost commercial and trading cities in Britain.

48 See: S. Gunn, "The middle class, modernity and the provincial city: Manchester, c. 1840–80", in *Gender, Civic Culture, and Consumerism: Middle-class Identity in Britain, 1800–1940*, eds Alan Kidd and David Nicholls (Manchester: Manchester University Press, 1999), 116–117.

A similar image emerges from films of one of the great engineering projects of the nineteenth century. The Manchester Ship Canal, which opened in 1894 and was filmed by the Warwick Trading Company and Mitchell and Kenyon in 1901, made Manchester, already the northern centre of the railway network, the fourth most important port in Britain. The city imported food and industrial raw materials from the colonies and exported textile goods, mostly cotton, to Australia, New Zealand, Canada, South Africa, China and India. Administered by the Ship Canal Company, the Manchester Ship Canal was built as an answer to the burden of port charges at Liverpool and the high rates charged by the railway companies for the carriage of freight between the two cities. The canal, which was intended to reduce the cost of transport, revived the cotton trade, and secured the city's regional role in the specialist trade of grain, meat, oil and fruit.[49] In 1901, both the Warwick Trading Company and Mitchell and Kenyon filmed the show given by a torpedo boat destroyer on the canal. Mitchell and Kenyon's *Torpedo Flotilla Visit to Manchester* and *Torpedo Boats Passing Through the Manchester Ship Canal* shows eight torpedo boat destroyers of the Royal Navy passing along the canal and under Barton Bridge. The latter also includes a scene recording the reception at the town hall in Manchester. A similar display of military might appeared in Warwick's *Torpedo Boat Destroyers* and *Through Manchester Ship Canal on a Torpedo Boat Destroyer*, a lost film apparently shot on board the boat itself. These films define Manchester as a modern urban space by displaying both the result of an impressive project such as the canal, and the application of technological development to the war industry in the form of warships.

Forty years before the opening of the canal, in 1853, Manchester had formally become a city defined, as Åsa Briggs suggests, by the cotton business as the trading centre of a region linked with the whole world. On the other hand, through operations like the building of the landmark Town Hall in 1877, Manchester's authorities displayed to the world its civic pride.[50] At the turn of the twentieth century industry and trade remained central

49 See Kidd, 103–120.

50 Briggs, 88–138.

to city life and its representation on film. The Town Hall often featured in official ceremonies recorded in early films such as Mitchell and Kenyon's *General Buller's Visit to Manchester* and in the opening shot of *Torpedo Boats Passing through the Manchester Ship Canal*. As Alan Kidd suggests, city planning in the second half of the nineteenth century saw an increasing emphasis on the political aspects of architecture and design. Manchester's monuments and statues, commercial and public buildings were covered with inscriptions celebrating civic, national and private pride.[51] The presence of monumental buildings like the Royal Exchange and the Town Hall or the great Victorian warehouses, as well as street improvements such as the widening of Market Street and the rebuilding of Deansgate reshaped the face of Manchester. In particular the great warehouses built with reinforced concrete and inspired by the architecture of Renaissance Italy gave Manchester a distinctive physical appearance and architecture that was regarded as the most innovative in the country. During the second half of the century the industrial and metropolitan character of the city, as Simon Gunn suggests, was the foundation of its modernity.[52] This cityscape can be seen in the visual record offered by early films at the end of a defining period in its history.

Manchester was nevertheless a socially-divided urban space. Prompted by Charles Booth's social research in the East End of London, in the late 1880s Fred Scott investigated poverty as an urban social problem in the Ancoats district of Manchester and in the Adelphi district of Salford. Scott's assessment of chronic poverty demonstrated that four-fifths of the population of Ancoats fell into the categories of the "poor" and the "very poor".[53] During the second half of the nineteenth century, the textile

51 See Kidd, 106.

52 Simon Gunn, *The Public Culture of the Victorian Middle Class: Ritual and Authority and the English Industrial City, 1840–1914* (Manchester: Manchester University Press, 2000), 10–13.

53 Fred Scott, "The Condition and Occupation of the People of Manchester and Salford", Transactions of the Manchester Statistical Society, 1888–1889, quoted in *Leisure, Gender and Poverty: Working Class Culture in Salford and Manchester, 1900–1939*, ed. Andrew Davis (Buckingham: Open University Press, 1992).

towns of Lancashire and the West Riding symbolized the respectability of a civilized working class that was in contrast with the physical degeneration characterizing life in the slums of London's East End. However, as Kidd has argued, a closer examination of Manchester reveals profound similarities with the semi-criminal underclass emerging from the investigations of London poverty exposed by Henry Mayhew in the 1850s and Charles Booth in the 1880s. Manchester had the worst pauperism in the country and this was caused, like in London, by irregular employment by the day or by the job. Consequently, the city was an unhealthy place in which to live, with an infant mortality rate of over 30 per cent well into the second half of the century. As in London, however, sanitary reform gradually brought improvement.[54]

The areas explored by Scott, like the East End of London investigated by Charles Booth, were mostly absent from representation in early actuality films. Mitchell and Kenyon's *Funeral in Manchester*, which was the only film shot in a working-class area of Manchester, is the only exception, showing modestly dressed men, children, and women lined on the side of the street. Policemen lead the parade of a coffin with flowers and a Union Flag on its top. As the flag suggests, this scene is likely to represent the return home for a young soldier who lost his life in South Africa. This is an image which greatly differs from films of the official return from the front of troops celebrated on the streets of the British cities. As a stage-managed cinematic space, Manchester at the turn of the twentieth century revealed more often the official face of urban life, with several films of military, imperial or royal rites and rituals. Indeed, as Simon Gunn has argued, events like royal and military visits or civic events had a highly stage-managed character which was aimed at making power visible in an orderly and theatrical manner:

> They developed in the context of a wider politics of urban space, encompassing the reconstruction of the provincial cities in the decade after mid-century and the creation of new ways of visualising identity and authority in public life. In this politics

54 See Kidd, 128–129. See also: Thomas S. Ashton, *Economic and Social Investigations in Manchester, 1833–1933: a Centenary History of the Manchester Statistical Society* (Hassocks: Harvester Press, 1977).

> of space, ritual and symbol were two of the key codes through which the middle class manifested its identity and defined its relations with the modern city. Ritual and symbol gave order to the freedom and flux of the city, and provided a means of containing the anxieties to which the conditions of urban modernity gave rise.[55]

The ritual and symbolic power of urban events in the age of high imperialism were given, in Manchester just as in the rest of the British Empire, a stage on the screen of the new-born cinematograph. Several films made by Mitchell and Kenyon contributed to give order to the freedom and flux of Manchester by filming ritual and symbol. Mitchell and Kenyon's *Manchester's Welcome to the Imperial Yeomanry* (1901), for example, shows a relatively small military parade at Exchange Station with soldiers in ceremonial outfits marching from left to right. The Imperial Yeomanry was a volunteer cavalry regiment and some of the soldiers are wearing colonial outfits, khaki tunics and hats with flattened sides. Factory chimneys are visible in the background. As Vanessa Toulmin suggests, the low turnout was probably due to the lack of real identification of the local population with the returning troops.[56] A much larger crowd can be seen in *Return of the Brave Manchester Volunteers*. This film was made in 1901 and recorded the return of troops from South Africa. Mitchell and Kenyon's *General Buller's visit to Manchester* was filmed that summer from the steps of the Town Hall. It shows empty horse-drawn carriages and mounted police passing through Albert Square with a huge crowd of spectators on the side of the road, and on trams with noticeable advertising for Lipton's tea and Mellin's food. General Buller, the former Boer War Commander-in-Chief, can be seen on a carriage outside the Town Hall with his wife and other dignitaries. Lord Roberts, who replaced Buller in South Africa as Commander, was also filmed during a visit to Manchester. In Mitchell and Kenyon's *Lord Roberts's Visit to Manchester* (1901), Roberts, the Lord Mayor and other civic dignitaries in ceremonial robes unveil a statue of Queen Victoria in front of a large crowd at the Piccadilly Gardens in the Northern Quarter. This film also shows mounted cavalry and police parading down

55 Gunn, "The middle class, modernity and the provincial city: Manchester, c. 1840–80", 123.

56 Toulmin, *Electric Edwardians: the Story of the Mitchell & Kenyon Collection*, 254–255.

a wide, crowd-lined street with factory chimneys in the background. *Royal Visit to Manchester, Owens College* was filmed by Mitchell and Kenyon on 12 March 1902; it shows the Prince and Princess of Wales (later George V and Queen Mary) and a number of mounted escorts and carriages emerging from Owen's College gateway. Civic pride was displayed in the images of the mounted police parades accompanying the police annual inspection recorded by Mitchell and Kenyon at the municipal Birchfields Park in south Manchester, which had been bought by the city council in 1887.[57] All these events focused on the royals and on war "heroes" such as Lord Roberts in a city whose inhabitants gathered to see them and whose built environment fittingly contained such rituals. Early films recorded the ritual power of these days and, especially when focusing on the visit of a Boer War fighter to the city, revealed an urban space profoundly connected with the larger empire.

Besides major national and imperial events, early films of Manchester record a variety of events reflecting different aspects of urban life, including sport and religion. Mitchell and Kenyon filmed football matches such as England v. Ireland at Manchester in 1904, and cycle races. During these events, elements and symbols of national pride were also displayed. In Mitchell and Kenyon's *Manchester and Salford Harriers Cyclists' Procession* (1901), Lord Frederick Roberts can be seen on a horse led by a man in a uniform as he doffs his hat and salutes the camera before being cheered by a small crowd previously behind the camera. Mitchell and Kenyon's *Manchester Wheelers' Annual Race Meet* (1901) features a three-team cycle race around a course marked by Union flags, the Royal Standard and crowds. *Race for the Muriatti Cup at the Manchester Wheelers' Annual Race Meet* (1901), which also features a cycle race, also shows a space dominated by the presence of the Union Flag. In *Manchester to Blackpool Road Races* (1903), Oxo advertising replaces the national flags. With a focus on icons of technological progress and nationalism – the Union Jack – these films thus reveal empire and trade as closely interconnected characteristics of urban life in Manchester.

57 *The Police Annual Inspection at Birchfields Park*, 1901, Mitchell & Kenyon.

Religious and local parades also defined the character of Manchester and its diversity at the turn of the century. *Congregation Leaving Manchester Cathedral* (1901) shows a glimpse of the cathedral with dressed-up men, women and a few children. Spiritualist and Catholic events were the subject of films by Mitchell and Kenyon.[58] The Spiritualist National Union was founded in 1901; in the same year Mitchell and Kenyon filmed *Manchester Spiritualists' Procession* in an urban park. Catholic Manchester is displayed in other films. In 1902, Mitchell and Kenyon filmed *Manchester Italian Catholic Procession* with a large crowd marching down a wide street with banners and crucifixes. In *Manchester Catholics' Whitsuntide Procession* (1904) Mitchell and Kenyon filmed a crowd-lined street at the base of a statue of Queen Victoria with a procession of children with well-dressed adults in attendance. The procession is headed by mounted police followed by clergy, a number of altar boys with banners, young girls in white dresses and veils, a marching boys' pipe band and men carrying a statue of Mary on a bed of flowers. A similar parade is shown in Mitchell and Kenyon's *Manchester Sunday Schools' Whit Monday Procession* (1904), which shows a procession marching on the tram tracks. In Mitchell and Kenyon's *Manchester Catholic Orphanage Boys* (1901) a large crowd watches a procession of a band, clergymen and uniformed boys with religious banners and followed by policemen. Mitchell and Kenyon's *Manchester Band of Hope Procession* (1901) represents the band marching down a crowd-lined street, young girls dressed in white; temperance organization working-class children with several banners promoting a healthy life with slogans such as "Drink no Wine", "Success to our Band of Hope", "Resist temptation resolutely" and "No beer in this child's home".[59] A similar parade also features in Mitchell and Kenyon's *Scenes in a Manchester Park* (1901), which shows a rather small parade followed by a horse-drawn tram, bicycles, people, and children posing for the camera. Late-Victorian and Edwardian

58 See Toulmin, *Electric Edwardians: the Story of the Mitchell & Kenyon Collection*, 166–169.

59 Andrew Prescott, "'We had fine banners': Street Processions in the Mitchell and Kenyon Films" in *The Lost World of Mitchell and Kenyon*, 125–126.

Manchester was thus a place of movement but also of order. The cinematic city expressed the regulated conduct of its inhabitants. This conduct was a reflection of the Manchester Municipal Code of 1893 and earlier regulations of market activities and street traffic bringing trams and cars to run along particular lines.

Early films of Manchester reveal a city charged with the capacity for free circulation expressed by images of wide streets allowing a maximum of free passage during peak hours; they show processions and other events, but not the alley and lanes behind these streets. Indeed, hardly any trace of the narrow alleys described by Engels in the passage from *The Condition of the Working Class in England* quoted earlier in this section can be found in early films.[60] On the contrary, early films reflect Stephen's view of Manchester as an industrial metropolis with a significant role in the establishment of Britain's rule among the nations of the world.[61] At the turn of the century, a contrast between wide streets and narrow alleys characterized the built environment: it was expressed by the development of Aldwych and Kingsway in London near the insalubrious lanes of the West End as much as by the slums of Manchester near central streets such as Deansgate and Shudehill. Filmmakers privileged the fluidity of movement in the wide thoroughfares of London, Liverpool, Manchester and, as I shall demonstrate in the next chapter, Glasgow.

60 Engels, 87–88.
61 Stephen, 580.

CHAPTER 4

Regional Urban Spaces: Scotland in Film

Where cargo ships, forests of pinnacles, an 135-foot tall dome and the familiar figure of Lord Roberts provide a complex image of Scotland in the Age of Empire.

During the second half of the nineteenth century, the steel industry replaced the iron industry as one of the dominant elements in Scottish economic life, and by 1900 over twenty per cent of British steel output was produced by Scottish firms.[1] The growth of heavy industry contributed to the continuing processes of urbanization and concentration of the population into relatively small and highly congested urban areas. In 1900, one-third of the population lived in Scotland's four major cities of Edinburgh, Aberdeen, Dundee and Glasgow.[2] At the turn of the century both these four cities and the beautiful scenery of rural Scotland attracted a number of film-makers. In this chapter, I examine early films of Scottish cities and draw particular attention to the most populated of them, Glasgow. The imperial project bound people of Britain together and the cinematic image of Glasgow presented themes equally expressed by films of Liverpool and Manchester. However, the cinematic image of Scotland in the 1900s also revealed characteristics specific to the country's identity, both in relation to industry and scenery.

1 See John F. McCaffrey, *Scotland in the Nineteenth Century* (Houndmills, Basingstoke: Macmillan, 1998), 84.

2 McCaffrey, 85.

Early cinematic representations of Scotland focused on images of urban spaces, which were filmed more often than the rest of the country, and rural landscapes. Films of rural areas include surviving scenes filmed in small villages and towns such as Elgin (which was filmed on the occasion of the opening of Cooper Park in 1903), Balmoral, Beith, Dumfries, Hawick, Greenock and Stranraer, as well as a phantom ride over the Tay Bridge in Fife. Parades to commemorate the coronation of George V in 1910 were filmed in both Annan and Broxburn.[3] In 1898, William Walker made a series of films in a variety of Scottish locations and explored different themes including military events and local processions.[4] Two years later, the Warwick Trading Company photographed from the express train on

3 Titles of the surviving films held at the Scottish Screen Archive and the British Film Institute: *Scenes at Balmoral* (1906, no credits available), *Street Procession in Annan* (1912, no credits available), *Workforce of Pollock's Caledonian Cabinet Works, Beith* (1901, Mitchell and Kenyon), *Coronation Parade in Broxburn*, 1910, Miller/Regal Cinema), *Dumfries Street Scenes* (1910, no credits available), *Opening Cooper Park, Elgin* (1903, no credits available), *Workforce of Scott & Co., Shipyard, Greenock* (1901, Mitchell and Kenyon), *Hawick Scenes* (1909–1910, no credits available), *Funeral of Provost Young, Stranraer* (1914, no credits available), *Railway Ride over the Tay Bridge* (1897, Peter Feathers), *Public School in Annan* (1901, Mitchell and Kenyon), *Congregation Leaving St Mary's Church, Dumfries* (1901, Mitchell and Kenyon), *Scholars Leaving St Andre's RC School, Dumfries* (1901, Mitchell and Kenyon), *Love Burn Street School, Dumfries* (1901, Mitchell and Kenyon), *Over the Tay Bridge on an Express Train* (1901, Mitchell and Kenyon). The Border Kinematograph Co., which was established in Hawick by local jeweller David Gaylor in 1899 and operated until 1906, produced a number of scenes filmed across the borders towns. Like Mitchell and Kenyon's, these were local films for local people and with a focus on a more rural Britain than the one emerging from most films of that period.

4 Titles: *Aberdeen Police Sports, Fire Light Hoarse, Barry and Buddon Camps, Rickarton Camp, The First A.V.A., First Volunteer Battalion Gordon Highlanders, Duttine Park Sports, Hospital Saturday Cycle and Trades Procession, Aberdeen South Breakwater in a Storm, Aberdeen Bathing Station Scenes, MacGregor's Gathering on the Clyde, Tramway Car Procession, Trains at Ballater Station, Banchory from the Railway, Scots Grey Entering Perth, Gordon Highlanders Battery Firing, Camerons at the Charge, Primrose League Gathering at Glen Tana, March of the Cameron Men, Lively Pillow Fight by Children, Curling and Skating at Glenmuick, Engines of the Metropolitan Fire Brigade.*

the Deeside Railway a film titled *Panorama of Banchory, Scotland.* In 1906, Charles Urban produced a long series of films called *Bonnie Scotland.* This series contained a few scenes filmed in Edinburgh, Aberdeen and Inverness, but it mostly consisted of a travelogue through rural areas and scenery such as the caves of Caithness and Fingal, Sutherland and Ross-shire, the Rogie Falls and Kyle of Lochalsh, St Kilda and the isle of Harris, Euston and Crewe, Rothesay and Obama, Iona, Oban and Loch Ness, Ayr and Balmoral, Loch Lomond and Lewis.[5] In the same year, Gaumont produced a shorter series of films shot in Hooghly, Rothesay Bay and Ayr.[6] The Scottish Highlands, already the subject of a large number of paintings, prints, etchings and photographs, were a favourite location for early filmmakers: arguably their wilderness attracted cameramen more consistently than any other British natural landscape. Charles Urban produced and distributed *Scottish Highlands in Winter*, a trip on the Caledonian Railway, in 1907, and, one year later, *A Trip through Scotland*, which included scenes filmed in the Highlands and also some urban views of Edinburgh and Glasgow. The Highlands also featured in B&C's *Bonnie Scotland* (1911), Kineto's *Through the Highlands* (1913), Claredon's *Scenes in the Western Highlands of Scotland* (1913), Gaumont's *Scottish Scenery* (1914) and Hepworth's *Scottish Shepherds at Work*, filmed by Florence Turner in 1914. Although most of these films are now assumed to be lost, what we know of them suggests they arose from a romantic nineteenth-century fascination with natural landscapes. Films of urban spaces, on the other hand, revealed a larger amount of information on the everyday life of people of Scotland in relation to

5 Titles: "Edinburgh", "London Scottish Volunteers", "Cave dwellers of Caithness", "Railway panorama through Sutherlandshire", "Across Ross-shire on the urban bioscope special", "Rogie falls salmon fishing", "From Kyle of Lochalsh", "From the mainland to St Kilda", "The isle of Harris and its tweed industry", "From Euston to Crewe", "From Crewe to Carlisle", "Rothesay to Oban", "Fingal's Cave", "Iona and the abbey ruins", "Oban to Loch Ness", "Round about Inverness", "Ayr: the Robbie Burns country", "Aberdeen and Balmoral", "Allen Water and Killicrankie", "The Trossachs and Loch Lomond", "Poverty Island: Lewis."

6 *Scotland* (Series), 1906, Gaumont: "A trip to the Hooghly", "Rothesay Bay", "A day in Burns' Land."

industrial dynamism, modernity, cultural life and empire. Beside Glasgow in film – which is the focus of the second half of this chapter – films of Edinburgh and Aberdeen seem particularly interesting for their focus on the cultural life of these cities.[7]

Edinburgh, the capital of Scotland, had a modest life on the screen with relatively few films of its streets and its buildings. Those that have survived are nevertheless interesting records of a city in constant expansion. The second half of the nineteenth century had seen great growth in the population, which doubled from 161,500 in 1851 to 317,000 in 1901. As Richard Rodger suggests, the development of branch offices and department stores in the city during years of commercial and financial expansion contributed to shaping a new city.[8] During the same period hundreds of new civic, public, governmental and ecclesiastical buildings were built in the city and they became an increasingly familiar part of the Victorian cityscape.[9] New markers were introduced in the mental landscapes of Edinburgh's population by means of slum clearance and public improvement schemes. As we already know, Charles Urban filmed Edinburgh twice in 1906 and 1908, but R.W. Paul was the first filmmaker to record moving images of

7 Relatively little evidence of early filmmaking in Dundee, the fourth city of Scotland, has survived and it is largely limited to three factory gate films, scenes of the fire brigade and a public funeral: *Employees Leaving Baxter's Jute Works, Dundee* (1901, Mitchell & Kenyon), *Employees Leaving Gilroy's* Jute Works, Dundee (1901, Mitchell & Kenyon), *Turn Out of the Dundee Fire Brigade* (1901, Mitchell and Kenyon), *Glimpse of Camperdown Work* (1912, Baillie John Noble/Oxford Picture House), *Funeral of Alexander Buist, Dundee* (1901, no credits). Arguably the most interesting example of early filmmaking in Dundee, *Strike in Dundee* was shot in December 1911 in the dockside area of Dundee around Crichton Street, now the site of the Caird Hall and City Square. Produced by Pathé as part of a newsreel, *Strike in Dundee* shows strikers trying to persuade blacklegs to join their ranks, and a large crowd in the city centre surrounding a carriage.

8 Richard Rodger, *The Transformation of Edinburgh: Land, Property and Trust in the Nineteenth Century* (Cambridge: Cambridge University Press, 2001), 487–491.

9 Brian Edwards and Paul Jenkins, *Edinburgh: the Making of a Capital City* (Edinburgh: Edinburgh University Press, 2005), 44–45.

streets of Edinburgh in 1896 and 1898.[10] In 1900, the Warwick Trading Company filmed a series of films that included scenes of people curling and skating in the city, troops boarding the train and the fire department.[11] In 1902, James Williamson filmed a series of films including a panorama of the city and scenes taken in Prince Street with views of the Old Town and Edinburgh Castle in the background.[12] In 1903, the Warwick Trading Company, Cecil Hepworth and Gaumont filmed the King and Queen's royal visit to Edinburgh, and their films included views of the arrival in Edinburgh and the procession from Dalkeith to Holyrood Palace.[13]

These films are now lost. Surviving films of Edinburgh before the First World War are very rare and include three films produced by Mitchell and Kenyon. *Lizars Edinburgh* (1901) and *Procession in Edinburgh* (1904) both feature troops parading on unidentified streets of the city. *Royal Scots Regiment at Edinburgh Castle* (1901) represents the most famous landmark in the landscape of the city, which still dominates its skyline: Edinburgh Castle. This film starts with a panning shot of the entrance to the Castle with soldiers on parade, and includes images of the entrance and its outer walls with double ranks of soldiers in Australian bush hats. Images of the Royal Scots Regiment contextualize the city of Edinburgh in a country, and an empire, at war in South Africa. Arguably the most interesting of the surviving films of Edinburgh, *Scottish National Exhibition* was filmed in 1908 by Gaumont and is a record of the Scottish National Pageant of

10 Titles: *Edinburgh*, *Street Scenes*.

11 Titles: *Winter Sport in Scotland*, *Scottish Sharpshooters Leaving Edinburgh*, *The Edinburgh Fire Department*.

12 *Scotland* (Series), including "Circular Panorama of Edinburgh", "Circular Panorama of the Forth Bridge" and "Panorama of Princes Street, Edinburgh."

13 Titles: *The Royal visit to Scotland* (Series), 1903, Hepworth ("Arrival of King and Queen at Edinburgh", "March past of Cameron Highlanders", "Scenes at Holyrood Palace", "Presentation of medals to black watch"); *The King's Visit to Edinburgh and Glasgow* (Series), 1903, Warwick Trading Co. ("King entering the city of Edinburgh", "King leaving municipal building, Glasgow"); *The King's State Visit to Scotland*, 1903, Gaumont (14 May) ("Procession from Dalkeith House to Holyrood Palace", "The King at Edinburgh Castle", "The King laying foundation stone, Glasgow Technical College").

Allegory, Myth and History. It opens with a view of the arched entrance to the Scottish National Exhibition grounds and a wide shot of crowds and people in smart clothes walking around exhibition displays and fairground amusements. Civic dignitaries and officials can also be seen on the exhibition grounds. The film also includes the remarkable presence of a group of African men, women and children performing dances in tribal dress. Their presence at the exhibition and on screen is a reminder of the ambition of Great Britain, and of the fascination of British people with exotic sights associated with the empire's colonies and dominions. Like films of the Zoological Gardens or the Crystal Palace in London, this film thus characterizes Edinburgh as a city with connections to the empire emphasized by the presence at the exhibition of subdued subjects. A local event was the subject of two surviving films made by unidentified filmmakers in 1911, when the Empire Palace Theatre in Nicolson Street caught fire during the act by German illusionist Sigmund Neuberger, known as The Great Lafayette, who died in the fire in the attempt to save his horse. *Theatre Fire in Edinburgh* and *Funeral of the Great Lafayette* record the burnt-out wreckage of the Empire Palace and the funeral of the artist but offer little insight into Edinburgh's urban life.

Like Edinburgh, Aberdeen was a city with a dynamic cultural life and its cinematic representation is a particularly interesting case study. During the nineteenth century Aberdeen became one of the great towns of Britain. Its port and the production of granite flourished. Its population increased from 60,992 in 1861 to 153,503 in 1901 and during the second half of the century major new streets such as Union Street and King Street and buildings like the Town House were developed.[14] A small group of early films of Aberdeen bear witness to the civic and national prestige of the city. Aberdeen briefly featured in the scenes filmed by William Walker in 1898 and in Urban's *Bonnie Scotland*. Walker's *Gordon Highlanders* shows

14 See Hugh MacKenzie, *The Third Statistical Account of Scotland: The City of Aberdeen* (Edinburgh: Oliver and Boyd, 1953). See also: Robert Mellor, "Aberdeen: the Great Century", John S. Smith and David Stevenson, *Aberdeen in the Nineteenth Century: the Making of the Modern City* (Aberdeen: Aberdeen University Press, 1988), 1–18.

a kilted pipe band leading a contingent of the Gordon Highlanders along Aberdeen's Union Terrace, before they moved to the army barracks in Edinburgh and eventually to South Africa. Once again, the Boer War brought distant places on the map of the British Empire into visual proximity. In 1913, when the Boer War had become a fading memory, Lord Roberts finally visited Aberdeen and an unknown filmmaker recorded the event. *Lord Roberts' Visit to Aberdeen* shows the former commander inspecting the Guard and on the pavement with local people. His carriage is shown driving along a crowd-lined street with mounted soldiers and policemen, and followed by a pipe band and the 4th Battalion Gordon Highlanders. The battalion is later presented the Colours in a large park. *Gordon Highlanders* was filmed with the background of the prestigious Marischal College, later described by John Betjeman as being "bigger than any cathedral, tower on tower, forests of pinnacles, a group of palatial buildings rivalled only by the Houses of Parliament."[15]

Between the making of *Gordon Highlanders* and the production of *Lord Roberts' Visit to Aberdeen*, two films were devoted to the University of Aberdeen and filmed during the ceremonies that took place to mark the 4th Centenary of the University. *Aberdeen University Quarter Centenary Celebrations* was produced in 1906 by Paul's Animatograph Works and represents a long procession of dignitaries in horse-drawn carriages. William Walker made a more complete film of this event.[16] His company, Walker's Royal Cinematograph, produced *Aberdeen* on 27 September 1906 when King Edward VII and Queen Alexandra visited the city to open the new extension to Marischal College in time for the fourth centenary of the university. This film opens with several views of the decorations around Union Bridge, Castel Street, Back Wynd, Union Terrace Gardens and the William Wallace statue, with His Majesty's Theatre and St Marks's Church in the background. The Town House on Castle Street, symbol of civic pride

15 Jeremy Betjeman, *First and Last Loves* (London: J. Murray, 1969), 28.

16 William Walker was a bookseller, magic lantern expert and pioneer Scottish filmmaker who produced hundreds of scenes between 1896 and 1911, and in 1898 was even invited by Queen Victoria to Balmoral Castle to exhibit his films.

just like the Town Hall in Manchester, can also be seen in the film. A large floral banner saying "Aberdeen University Quarter Centenary Celebrations September 1906" is situated on Union Bridge and behind the statue of King Albert. A second banner saying "God Bless Our King And Queen" covers the front of St Nicholas Cemetery on Union Street. Band of Royal Scots, Fifeshire and Forfarshire Imperial Yeomanry, and Seaforth Highlanders can be seen parading on the streets of Aberdeen. Queen Alexandra and King Edward VII are filmed as they arrive at Holburn Street Station and are escorted with a number of dignitaries through the crowded streets to the entrance of Marischal College on Broad Street. These few surviving films define Aberdeen as an important cultural and military urban centre on the map of Britain: Walker's *Aberdeen*, in particular, reveals a monumental city fit to receive the King and thus acquire a certain importance on the map of the British Empire. However, imperial centrality and industrial might in Scotland were more consistently expressed in films of Glasgow.

Glasgow: Films of an industrial city in expansion

As Brian Dicks suggests, the title of "shock city" north of the border fell to Glasgow, where economic vitality and social decay in the slums went, as in Manchester, Liverpool and London, hand in hand.[17] The earliest moving pictures of the city had been filmed by R.W. Paul in 1896 and consisted of a series of views of streets in Glasgow's central district, including films of Argyle Street, Buchanan Street, Sauchiehall Street and Rothesay Pier.[18] These films are now lost as are those produced by Charles Urban in 1906

17 Brian Dicks, "Choice and constraint: further perspectives on socio-residential segregation in nineteenth-century Glasgow with particular reference to its West End", in *Perspectives of the Scottish City*, ed. George Gordon (Aberdeen: Aberdeen University Press, 1985), 91–124.

18 Title: *Glasgow* (Series: "Buchanan Street", "Sauchiehall Street", "Rothesay Pier", "Argyle Street").

and 1908, and Gaumont's *Glasgow Sports*, filmed in 1903 and featuring highland dancing, wrestling and jumping. *Macmillan's Shop front*, which was filmed in 1905, is one of the earliest films of Glasgow to have survived: it represents a popular tea room at 151 Union Street and briefly shows the wealthy central district of Glasgow.[19] These early films of Glasgow reveal the imperial connections of the city and its industrial power, while also offering a glimpse of civic pride, municipal socialism and its achievements. Cinematic Glasgow confirms themes which also characterized English cities, on the one hand, and expresses different cultural and social dynamics on the other.

John MacKenzie suggests that late Victorian Glasgow was one of the best commercially and politically-connected cities of Britain.[20] From the 1790s it became a major centre for cotton processing and, with the canalization of the river Clyde, large vessels were able to reach right into the heart of the city. By the mid nineteenth century, heavy engineering and shipbuilding became increasingly important and a number of shipping lines were founded. As MacKenzie suggests, these processes brought transformations into the built environment and, by the end of the century, Glasgow was a amalgamation of an industrial cities like Manchester and shipbuilding centres like Liverpool.[21] This economic development was accompanied by increasing social problems. Between 1841 and the outbreak of the First World War, Glasgow quadrupled its population and grew tenfold from the 1,864 acres of 1830 to 19,183 acres in 1912. In 1901, 761,709 people, 17 per cent of the Scottish population, lived in Glasgow and this resulted in overcrowded rooms in masses of overcrowded homes in the tenement structures. Throughout the nineteenth century, the city also absorbed the surrounding ring of industrial and residential burghs and led the growth of the Scottish economy. Thanks to a number of industrial inventions and technical developments in the steel industry, the booming needs of the

19 No credits survive.

20 John M. MacKenzie, "'The Second City of the Empire': Glasgow – imperial municipality", in *Imperial Cities: Landscape, Display and Identity*, 215–220.

21 MacKenzie, 218.

empire's economy in Glasgow began to draw in labour and Glasgow was transformed from a textile and building centre into a major shipbuilding and marine engineering centre.[22]

This was a handsome city characterized by its industries, but also by the neoclassical design of its buildings on Bath Street, West Regent Street, Argyle Street, Garden Square, and Queen's Crescent. On the other hand, social problems associated with the slums, overcrowding, public health and alcohol consumption also characterized the city.[23] Just as the working-class districts of Manchester had been photographed by Samuel Coulthurst in the 1890s and the poor of London by John Thompson in the 1870s, the slums of Glasgow found expression in the photographs taken by Thomas Annan for Glasgow Corporation.[24] In her investigation of the representation of the city in the visual arts, Caroline Arscott has explained that the haunting images of disease-ridden crumbling alleyways photographed by Annan rendered in an arresting form those closely spaced walls that signified overcrowding, foul air, sewage and disease.[25] On the contrary, early films focused on the wealth of Glasgow and, through selection and omission of urban subjects, gave a prosperous image of the city. Just as in Manchester and Liverpool, official events were consistently filmed on Glasgow's streets. In 1903, an unidentified filmmaker made a film of King Edward VII and Queen Alexandra laying the foundation stone for the Glasgow and West

22 See Andrew Gibb, *Glasgow: the Making of a City* (London: Croom Helm, 1983), 113. See also: Nick F. Morgan, "Building the city", in *Glasgow: 1830 to 1912*, vol. II, eds W. Hamish Fraser and Irena Maver (Manchester and New York: Manchester University Press, 1996), 9–10. See also Peter Hall, *Cities in Civilization: Culture, Innovation, and Urban Order* (London: Phoenix Giant, 1999), and Smith, "Class, skill and sectarianism in Glasgow and Liverpool, 1880–1914."

23 See David Daiches, *Glasgow* (London: André Deutsch, 1977), 158–166.

24 See Thomas Annan, *Old Closes and Streets of Glasgow, 1868/1877* (Glasgow: City Improvement Trust, 1879), and Thomas Annan, *Photographs of the Old Closes and Streets: a Series of Photogravures, 1868–1899* (Glasgow: James Maclehose, 1900).

25 Caroline Arscott, "The representation of the city in the visual arts", in *The Cambridge Urban History of Britain*, vol. II and III, ed. Peter Clark (Cambridge: Cambridge University Press), 823.

of Scotland Technical College in the presence of a large crowd.[26] Formed in 1887 following the 1882 Educational Endowments Act, this college was situated at 138 George Street. Its increasing number of students required larger premises. In this film the King enthusiastically lays the stone of what was to be, upon completion in 1912, one of the largest educational complexes in Europe. Like the films made in Aberdeen in 1906, these scenes emphasize a cultural dimension of city life. Three years later, militarism in the form of boys' troops marching orderly and led by soldiers on horseback is shown in Mitchell and Kenyon's *Boys' Brigade Inspection at Yorkhill*, filmed in Kelvingrove Park on the River Kelvin in the West End of the city.

Urban grandeur and social degradation, as MacKenzie suggests, contributed to the creation of an imperial self-image which combined wealth and poverty, and whose social weaknesses found an answer in local government-led social reforms, defined as municipal socialism, on a great scale. Glasgow's imperial image was to be found in the powerful iconography of empire developed by Glasgow's manufacturers in producing and advertising their goods – for example, Camp Coffee and Tennent's Indian Pale Ale – and in the name of its streets: Virginia, Savannah, Jamaica, Tobago and West Nile were all names of streets in the successive grids forming modern Glasgow. The imperial theme, MacKenzie continues, also emerged from examples of orientalism in architecture – for example, the Egyptian and Hindu elements in Alexander Thompson's St Vincent Street and Queen's Cross churches – and the statues of national heroes like Colin Campbell, Lord Clyde, and David Livingstone; the Doulton Fountain on Glasgow Green contained sculptures of Canada, India, South Africa and Australia, as well as British soldiers, Queen Victoria and imperial lion masks.[27] Imperial iconography was also to be found in early moving pictures of the city. Glasgow, like Liverpool, was connected to the empire through its port, which was the subject of at least two films. *Photographic Convention*, film in 1898 by an unidentified filmmaker, is the earliest surviving film of Glasgow

26 Title: *Edward VII Lays Foundation Stone*.

27 MacKenzie, "'The Second City of the Empire': Glasgow – imperial municipality", 218, 223–226.

and it shows members of the photographic convention carrying tripods and cameras on the deck of the Clyde steamer *Lady Rowena* on her arrival at Glasgow. The dockyards were central to the economic life of the city and, eight years later, Mitchell and Kenyon filmed *SS Mongolia Leaving the Clyde*, which includes scenes of passengers boarding the ship and waving from the decks to a very large crowd on the dockside. The SS Mongolia was a passenger-and-cargo liner built for Pacific Mail Steamship Company in 1904. This film reveals the role played by the city in the emigration to the United States: like Liverpool and Southampton, Glasgow saw a large number of people migrating to the States in search of better economic and social opportunities.

Icons of imperial culture were to be found in the exhibitions that took place in the city. The international exhibitions of 1888 and 1901, as Maurice Lindsay suggests, flowered out of the city's sense of prosperity and the belief that Glasgow had something to show off to the world. Glasgow's successful series of major shows demonstrated the cosmopolitanism of the city and, by exploring everyday life in relation to art and industry, they powerfully emphasized the connections between the city and the empire.[28] The 1901 exhibition was the most extensive and ambitious the city had ever seen and it had the purpose of demonstrating the development of art, industry and science which characterized nineteenth-century Glasgow. It also displayed innovations like the motor-car and several exotic displays of other cultures. It was thus described in *The Times*:

> The site is, as in 1888, Kelvin-grove park. Besides numerous pavilions in the grounds there are three main buildings, the industrial, machinery, and grand concert halls. The first of these, which, attached to the art gallery dominates the design, in 7000ft. by 360ft., and covers about six acres. Its most striking external feature is a great central dome whose apex is 135ft. from the ground.[29]

28 Maurice Lindsay, *Portrait of Glasgow* (London: Robert Hale & Company, 1972), 151–153.

29 *The Times*, Jan 21; p. 6.

This exhibition, which marked the peak of the city's economic prosperity, also had strongly orientalist elements. For example, it displayed four "triumph and navigation" sculptures celebrating the conquest of distant territories of the British Empire and, more broadly, it was dominated by exhibits from India and many other territories, including the recently colonized Rhodesia.[30] William Dickson filmed for Biograph Syndicate the opening ceremony attended by the Duke and Duchess of Fife and a panorama of the exhibition. These films are now lost but their themes suggest the display of Glasgow as a city with close connections to the empire and an example of urban modernity.[31]

Compared to films of London, however, the imperial theme emerges less prominently from cinematic Glasgow. The city was also a powerful municipality which employed public ownership of services to mitigate the profound social problems affecting it. A glimpse of local government-led social reforms in Glasgow can be observed in early films of the city. In 1890 and 1894, Glasgow brought electricity and the tramway system under public ownership to tackle its social problems and with the aim of improving the lives of its people.[32] The sight of public trams, as early films recorded, dominated life on the streets of Glasgow. As they represent both city dwellers wearing working class clothes and richer people, early streets scenes reveal a dynamic city where public transport allows its central districts to be experienced by all classes. *Glasgow and the Clyde Coast*, produced by Kineto in 1909, represents holidaymakers travelling by train and steamer down the coast from Glasgow to Rothesay and Ettrick Bay, and it includes a phantom ride filmed from a tram in Sauchiehall Street. The film opens with a shot of the City Chambers in George Square with pedestrians strolling around the municipal building. This is followed by a tracking shot from a tram travelling east along Sauchiehall Street, busy

30 See MacKenzie, "'The Second City of the Empire': Glasgow – imperial municipality", 226–228.

31 Titles: *Opening of Glasgow International Exhibition Panorama of Exhibition*.

32 W. Hamish Fraser, "Municipal Socialism and Social Policy", *The Victorian City: a Reader in British Urban History: 1820–1914*, eds Robert J. Morris and Richard Rodger (London and New York: Longman, 1993), 258–280. See also McCaffrey, 86–88.

with trams, horses and carts, from Rose Street to Jamaica Bridge and across the river Clyde. At this point, the filmmaker leaves Glasgow following the route taken by holidaymakers and follows them to their destination, Rothesay. Mitchell and Kenyon's *Glasgow Trams* (1901) shows an open-topped tram turning the corner from St Vincent Street into Hope Street, and continues with a phantom ride filmed from a tram on Argyle Street and Hope Street: the streets are filled with people, other publicly-owned trams travelling in the direction of Queen's Park, Govan, Shawlands and Shieldhall, and horse-drawn carts. The work of Mitchell and Kenyon also offers a visual record of classical style architecture on Argyle Street and Hope Street with a glimpse of their mid-century iron-framed and stone warehouses.[33]

At the end of August 1900, an outbreak of bubonic plague was reported in Glasgow.[34] Despite sanitary reforms and improved living conditions, the quality of life in the poorest districts of the city was still low. Nevertheless, the spread of the disease was contained with only twenty-four reported cases by 22 September, and the provisions of the Infectious Diseases Notification Act quickly stamped out the disease.[35] The city became a more salubrious place where to live and Glasgow's confident urban modernity was reflected in films such as those made by Mitchell and Kenyon. *Jamaica Street Glasgow* (1901) is a good example of this tendency. It was filmed in the city centre from a fixed camera position. Jamaica Street appears as a very busy street with pedestrians, a variety of horse-drawn carts laden with barrels and wood, Glasgow Corporation trams and several carriages. The first half of *Jamaica Street Glasgow* is filmed looking south with Howard Street turning to the left. The street is wide, with dense traffic in both directions: several carts enter the frame diagonally from the right and exit to the left, from the background to the foreground. On the other side of the street, several trams can be seen as they enter the frame from the left and exit from the right, in the background. The second half of *Jamaica Street Glasgow*

33 See Daiches, *Glasgow*, 163.

34 "Plague at Glasgow", *The Times*, Friday, Aug 31, 1900; p. 6.

35 "The outbreak of plague at Glasgow", *The Times*, Saturday, Sep 22, 1900; p. 9.

is filmed further down on the other side of the road looking north, with several carts turning left to Clyde Street under the railway bridge. In both shots, a large number of men, women, and children walk down Jamaica Street, wearing either working clothes or elegant dresses, pursuing their daily activities. Like in Liverpool or Manchester, the central districts of Glasgow appear rational with a ordered stream of vehicles in line and a clear distinction between roadway and pavement: the cityscape is thus characterized by movement and fluidity, and by order within them. More broadly, Glasgow appears in early films like *Jamaica Street Glasgow* as a dynamic urban space that seems to confirm the claim made by historians like John K. McDowell, J.B. Gilfillan and Charles Oakley that Glasgow was the second city of the British Empire.[36] As films of her cities thus exemplify, Scotland is displayed on film as a stable part of the British Empire and the Union, and yet animated by regional cultural dynamics.

War, peace and industry

Response of the Glasgow Tramway Men to the Country's "Call to Arms" was filmed on 7 September 1914 in front of the City Chambers of Glasgow in George Square and shows tramway men forming the 1st Battalion in response to the "call to arms." The First World War was about to open an entirely new dimension to non-fiction filmmaking, war technology and urban iconography in the West. By the mid-1910s, the acceleration of the technological progress was applied to war on a larger scale than ever before. The Boer War, which had dominated the cinematic landscape at the turn of the century, was by now a fading memory. The war in South Africa had

36 See: John K. McDowell, and John B.S. Gilfillan, *The People's History of Glasgow: An Encyclopedic Record of the City from the Prehistoric Period to the Present Day* (Glasgow and London: Hay Nisbet, 1899); Charles A. Oakley, *The Second City: the Story of Glasgow* (Glasgow and London: Blackie, 1946).

challenged Britain's confidence in the real or imaginary civilizing role of her empire. It also caused introspection, doubts and vicissitudes about the imperial project: films ignored this dimension of the war as unambiguous images of hero-worship unfolded on the cinematic screen.

The war effort in South Africa is one of the themes linking the cinematic landscapes of Britain at the turn of the century. Other common themes, such as empire and industrial modernity, emerge from this study. Films recording an increasing ease of movement in the city and a focus on monumental landmarks, exhibitions and orderly traffic in city centres thus suggest a thematic continuity between the cinematic image of London and other British cities. During the nineteenth century streets were widened, and modern lighting, sewage systems and water supplies created in most British cities. Modern developments thus affected life in cities at times reluctant to embrace modernity, and where pockets of neglect were not far from the most recent urban development. Urban life was also affected by sectarian violence, which was on the rise both in Liverpool and Glasgow.[37] Filmmakers contributed to the image of urban modernity and imperial splendour by ignoring slums and urban unruliness in films of Manchester, Liverpool or Glasgow. From the top of a moving tram or fixed in front of a lively street scene, the camera captures the fluidity constituted by spatial practices in the city centres and experienced by those walking their wide, well-lit and clean streets.

An understanding of these cinematic cities makes explicit the dichotomy between images of everyday urban life and celebrations represented in early films produced by both local and national pioneer filmmakers. In particular, it provides a reassessment for the meanings revealed by events like parades and other celebrations. The popularity of these rites and the expansion of popular visual culture went hand-in-hand, and contributed to the emergence of a national language of ritual activity related to early films. Films of late Victorian and Edwardian Britain thus reveal nationalism as an uneven sense of belonging to a common group defined by race, language, history and territory, and as celebration of the state-nation as

37 Smith, "Class, skill and sectarianism in Glasgow and Liverpool, 1880–1914", 157–167.

the culmination of the group. The enormous vitality of rites related to nationalist feelings was the subject of several films recording the opulence of national celebrations including St George's Day and Trafalgar Day. Royalist ideology, on the other hand, was expressed by films of royal visits to urban centres outside London. The British monarch was also the emperor or empress of India and consequently royal events in British cities were both royal and imperial celebrations. Ultimately, films contributed to bringing to the national stage recorded images of urban life defined by imperialist, nationalist and royalist ideology, as well as civic pride.

There is a clear sense of economic invention and mercantile dynamism emerging from films of the industrial cities investigated in this and the previous chapter. The growth of the industrial and mercantile power of Britain was, from the end of the eighteenth century, accompanied by global expansion. In Victorian and Edwardian Britain, a consensus of nineteenth-century capitalism reflected British expansion to the far reaches of the globe. In the films discussed here, images of industrial and mercantile dynamism thus express another face of British imperialism. Images of large-scale operations at the port of Liverpool or the ship canal in Manchester reveal the technological advancement of industrial Britain more consistently than films of London, which were more concerned with political, royal and financial aspects of urban life. In films of urban centres outside London, there also seems to be more consistent evidence of local government-led social reforms expressed, for example, by a more direct focus on recurring images of publicly-owned means of transportation and civic buildings. In films of London, this evidence mostly emerges from films of life in suburbs such as Ealing. A complex series of themes and representational strategies also emerged from films made in a territory characterized by a hybrid location in the imperial system: Ireland.

CHAPTER 5

Hybrid Space: The Irish City

Where the author raises questions on the place of Ireland in the imperial system and we meet a character from Joyce's *Ulysses* outside the Church of Francis Xavier in Dublin.

The Boer War had a diverse impact on British culture and society. It was portrayed in the media with jingoistic undertones and yet undermined people's faith in the imperial adventure. For example, the events in South Africa prompted Reginald Brabazon, the twelfth Earl of Meath, to question the future and survival of the British Empire.[1] In previous chapters we have observed how the war became a recurring theme in early British films, and the cinematic image of Ireland was also affected by the conflict against the Dutch settlers. In 1902, Mitchell and Kenyon produced *Regiments Returned from the Boer War to Victoria Barracks, Cork*, which represented the Royal Munster Fusiliers, an Irish Infantry Regiment of the British Army, marching to the barracks (later renamed for Cork-born and first commander-in-chief of the Irish Free State Michael Collins). Scenes like this were common in films made in Britain and formally Mitchell and Kenyon's film does not differ from those shot on the other side of the Irish Sea. However, the perception of the war in Ireland was very different from that in Britain. As Keith Jeffery suggests, the Boer War represented the unacceptable face of the empire, a violent challenge for power and conquest which was popu-

1 Meath, "British youth and the Empire", in *King and Country*, London (1902): 23 (Meath Archive). Quoted in David H. Hume, "Empire Day in Ireland, 1896–1962", in *An Irish Empire? Aspects of Ireland and the British Empire*, ed. Keith Jeffery (Manchester: Manchester University Press, 1996), 149.

lar in Britain and loathed in Ireland, where the nationalist movement was invigorated by the opposition to the war and attempted to dissuade Irishmen from joining the British army.[2] An Irish "Transvaal Committee" was formed to support the Boer republics and more broadly, as Robert F. Foster has argued, the involvement of Britain in the Boer War "focused much moderate Irish opinion into an anti-imperial mould, and provided a mobilizing cause against the government."[3] As Charles Townshend suggests, there was also a nationalist reaction to the violent jingoism of the war.[4] It is remarkable therefore that a film such as *Regiments Returned from the Boer War to Victoria Barracks, Cork* represents an Irish landscape as an unproblematic and settled part of the British Empire.

Between 1841 and 1901, while the population of Britain more than doubled, the Irish population almost halved from eight million to four and a half million because of famine and emigration.[5] The economic misery of Ireland contributed to exasperating the anti-imperial mould which grew stronger in the early twentieth century. However, no trace of this emerged from early films of urban and rural life in that country, and Irish nationalism was structurally absent from representation. Cities and landscapes of Ireland in early films offered an uncomplicated vision of the country as a part of the union with Britain as well as a source of a very special kind of exoticism. Edward Said, who addressed Ireland in the context of imperialism and colonialism, put forward parallels between Orientalism and Irish Celticism arguing that the image of Ireland resulted from a history, geography, culture and population written and represented by what the British said about them.[6] Similarly, as Jeffrey Richards suggests, the cinematic image

2 Keith Jeffery, "The Irish military tradition and the British Empire", in *An Irish Empire? Aspects of Ireland and the British Empire*, 95–96.

3 Robert F. Foster, *Modern Ireland: 1600–1972* (London: Penguin Books, 1989), 433.

4 Charles Townshend, *Ireland: the 20th century* (London: Arnold; New York: Oxford University Press, 1999).

5 Stephen Howe, *Ireland and Empire: Colonial Legacies in Irish History and Culture* (Oxford: Oxford University Press, 2000), 22–24.

6 Edward W. Said, "Reflections on Ireland and Postcolonialism", in *Ireland and Postcolonial Theory*, eds Clare Carroll and Patricia King (Notre Dame, IN, and Cork, Ireland: University of Notre Dame Press and Cork University Press, 2003), 177–185.

of Ireland throughout the twentieth century was largely in the hands of American and British film industries and resulted in a cultural construction in which Irish men and women had extensively collaborated.[7] Early actuality films of Ireland also offered a representation of the country as seen through the eyes of the British colonizer: at the turn of the twentieth century, with no relevant film production to speak of, cinematic images of Ireland mostly came from the work of British filmmakers travelling and filming in the country. Early cinematic Ireland thus reflects the country's status as a colony and a part of the Union with Great Britain. And yet, films of Irish cities also hold images of national Irish culture, such as its Catholic congregations, which emerge from a broader view of the country as part of the Union in a hybrid result which reflects much of the complex activity of placing modern Ireland within the British imperial system.

This hybridity raises the question of how to view Ireland's place in the empire. Keith Jeffery, among other historians, has argued that Ireland was paradoxically both imperial and colonial, and its people both sustained the British imperial system, as soldiers, settlers and administrators, and undermined it as rebels against it.[8] Consequently, a sound understanding of early filmmaking in Ireland cannot neglect the questions raised by the "unusual" Union of the kingdoms of Ireland and Great Britain. David Fitzpatrick has explained that that union was a response to the Irish rebellion of 1798, which was organized by the republican revolutionary group the United Irishmen, and masked a hybrid administration with manifest colonial elements of occupation: Ireland was thus allowed substantial representation in both Houses of Parliament in London, while its administration by direct rule under a "Lord-Lieutenant" or "Viceroy" remained distinctively colonial in both form and function.[9] The Irish Viceroy provided the proconsul prototype for the later regime established in India and, on

7 Jeffrey Richards, "Ireland, the Empire and film", in *An Irish Empire? Aspects of Ireland and the British Empire*, 25.

8 Keith Jeffrey, "Introduction", in *An Irish Empire? Aspects of Ireland and the British Empire*, 1.

9 David Fitzpatrick, "Ireland and the Empire", in *The Oxford History of the British Empire: The Nineteenth Century*, vol. III, ed. Andrew Porter (Oxford and New York: Oxford University Press, 1999), 494–498.

the other hand, the presence of Irish MPs in the British Parliament, such as nationalist Charles Stewart Parnell, resulted in recurring condemnations of colonialism from the Westminster benches.[10] Historians such as Alvin Jackson have also pointed out that the occupation of high offices by Irishmen such as Hugh Cairns, Edwards Gibson, Lord O'Hagan and Lord Russell of Killowen in successive British governments was complemented and contradicted by colonial political functions embodied in the Viceroy.[11] Fitzpatrick has summarized this hybridity well:

> The imprint of Ireland may be detected in virtually every colonial institution, ranging from schools and police forces to land law, fraternities, political parties, and the churches. Likewise, the imprint of Britain may be found in every Irish institution, signifying the ambiguity of Ireland's location in the Empire. Through the exercise of imagination, the nineteenth-century Irish might elect to play the parts of colonials (whether resentful or deferential), metropolitans, or colonizers. To be Irish was, among other things, to face that unsettling choice.[12]

Early British films of Ireland also presented hybrid forms of representation and themes, with elements of both colonial filmmaking and national film practices.

Dublin's first screenings took place on 17 April 1896 at the Erin Theatre of Varieties, when a number of Lumière films such as *Arrival of a Train at La Ciotat* and Alexandre Promio's actuality films shot in London and representing the imperial landmarks of the city were exhibited.[13] Nevertheless, in 1896 the Irish paid little attention to the new invention and, as Denis Condon reports, newspapers offered more information on early radiography than on early films.[14] During the following year, when Alexandre

10 These are well described in: Howe, 42–50.

11 Alvin Jackson, "Ireland, the Union, and the Empire, 1800–1960", in *Ireland and the British Empire*, ed. Kevin Kenny (Oxford: Oxford University Press, 2004), 125.

12 Fitzpatrick, 520.

13 See Kevin Rockett, Luke Gibbons and John Hill, *Cinema and Ireland*, London: Routledge, 1987.

14 Denis Condon, "Spleen of a cabinet minister at work: exhibiting X-ray and the cinematograph in Ireland, 1896", in *Film History and National Cinema*, eds John Hill and Kevin Rockett (Dublin: Four Courts, 2005), 70.

Promio filmed the cities and landscapes of Ireland, the first moving pictures of the country increased popular interest in cinematography. Promio's Irish films depicted predominantly urban spaces and were taken in Dublin and Belfast. His films of Dublin included subjects such as the 13th Hussars, a cavalry regiment of the British Army, parading through the streets of the city (*Hussars Passing Through Dublin*). Promio also filmed the inevitable departure of a train at Amiens Street Station, presently Connolly Station, and a street scene on Sackville Bridge, now O'Connell Bridge, which joins Dublin's main thoroughfare Sackville Street, now O'Connell Street, to D'Olier Street. This film was taken from the bridge looking northwards, and the imperial sight of Nelson's Pillar, a 134 ft tall British military memorial, appears in the background. Notwithstanding the sight of Nelson's Pillar, Promio brought a change in the representation of the city by filming ordinary people, pedestrians, cyclists and horse-drawn vehicles when most nineteenth-century views of Sackville Street in paintings, prints and photography, as Harvey O'Brien has argued, enhanced its colonial grandeur with a focus on depicting the city as a landscape without living people.[15] Promio also filmed Grafton Street, one of the principal shopping streets and cross-city routes in Dublin city centre, in *Crowds Watch City Fire Brigade*: this film shows horse-drawn fire engines dashing along Grafton Street followed by a large crowd of people. In Belfast, Promio filmed a panorama of the city including Queen's Bridge, Ann Street and Castle Place. As Kevin Rockett has argued, from the early 1900s exhibition of largely foreign-produced films of Ireland began to grow.[16] Promio's work set an example for other filmmakers, as he was followed soon by other cameramen filming life in Ireland for the Warwick Trading Company, Cecil Hepworth, R.W. Paul, Kineto, Charles Urban, Northern Photographic Works and Biograph Syndicate.

Pioneer cameramen filmed Ireland with selective gaze on its political geography: films of Irish cities and the countryside portrayed an attractive

15 Harvey O'Brien, *The Real Ireland: the Evolution of Ireland in Documentary Film* (Manchester University Press, 2004), 21–23.

16 Kevin Rockett, *Cinema and Ireland* (London: Croom Helm, 1987), 3–6.

country at rest, with busy modern urban spaces and picturesque scenery. As O'Brien has argued, Alexandre Promio produced moving pictures of Ireland as a country lacking unique features which would single it out from the rest of the United Kingdom and, like in his representations of London and Liverpool, he filmed Ireland as a modern and urbanized country.[17] At least in part, a similar image of Ireland appears in the films produced by Sagar Mitchell and James Kenyon, the most striking surviving examples of British filmmaking in Ireland for variety of subjects and technical quality. These are over twenty films made between May 1901 and December 1902 in association with three travelling film exhibitors: the North American Animated Photo Company, the Thomas Edison Animated Photo Company and the fairground showman George Green.[18]

In terms of subjects and themes, early travel films of Ireland such as those produced by Hepworth and Urban remind one of the travelogues filmed during the same years in the colonies of the British Empire. As much as the films of the colonies which will be discussed in the following chapters, Hepworth and Urban's actuality films were exhibited for British audiences and emphasized picturesque and "typical" scenes of life in the country. On the contrary, the films made by Mitchell and Kenyon belong to the group of films advertised at that time as "Local films for local people", and were taken during the day in a certain place and shown on the same evening in the same city or town.[19] As we have seen in Chapters 3 and 4, Mitchell and Kenyon produced most of these "local films" in northern England. Ireland was the only non-British territory visited by Mitchell and Kenyon, and their films, rather than being aimed at British audiences, were shown to local people in the towns and cities where the films were shot. Their local topical films were based on the idea of local audiences which would have gone to see these films in order to recognize the places were they lived and even themselves on the screen: slogans such as "See yourselves as others see

17 O'Brien, 21–30.

18 See Robert Monks, "The Irish Films in the Mitchell and Kenyon Collection", in *The Lost World of Mitchell and Kenyon: Edwardian Britain on Film*, 93–102.

19 See Vanessa Toulmin, *Electric Edwardians: the Story of the Mitchell & Kenyon Collection* (London: BFI Publishing, 2006).

you" were used to promote screenings in fairground tents or local meeting halls and music halls. In 1903, an article published in the *Sligo Champion* described how Irish audiences were amused by these films which caused "rounds of laughter as those in the audience recognized themselves or their neighbours in the living pictures."[20] Both the *local* films made by Mitchell and Kenyon and the *colonial* films made by other filmmakers in Ireland contained a series of sharp contrasts reflecting the role of Ireland in the British imaginary, and embodied in the dichotomy between peasant traditions and urban modernity. A number of films taken in Ireland between 1897 and 1914 focus on picturesque subjects from the rural parts of the country, reminiscent of earlier postcards and photographs representing subjects such as Irish colleens at work. *Typical Scenes in Ireland*, produced by the Warwick Trading Company in 1904, *Irish Life and Character* and *Irish Scenes and Types*, both produced by Charles Urban in 1906, *In Ireland: Excursion to Killarney* (British Pathé, 1908), *A Cattle Drive in County Galway* (R.W. Paul, 1908), *An Irish Stag Hunt* (Warwick Trading Company, 1902) and *Irish Life: Pig Raising* (General, 1913) represent everyday activities in the Irish countryside such as the farming practices and the management of livestock. Irish urban spaces were the other main focus of pioneer filmmakers and these films, which also surpassed their rural counterpart in number, offered a less stereotypical view of Ireland. At the turn of the twentieth century Belfast, Cork and Dublin were the three largest urban areas in Ireland and the actuality films which represented their streets, inhabitants and forms of transport offer an image of dynamic and modern cityscapes. Smaller centres such as Derry and Wexford emerged in their cinematic image as poorer and more "rural" cityscapes (there is no record of films representing other big towns such as Galway or Limerick in this period). In what follows, I mainly focus on the three largest cities and their cinematic representation and, in particular, on the work of Mitchell

20 *Sligo Champion*, 17 January 1903: 5; quoted in Niamh McCole, "The cinematograph in provincial Ireland 1896–1906: exhibition and reception", in *National Cinemas and World Cinema*, eds John Hill and Kevin Rockett (Dublin and Portland, Oregon: Four Courts, 2006), 35.

and Kenyon. I investigate the emergence of imperial themes in films of Dublin, Belfast and Cork for, as Jeffery suggests, imperial resonance was to be found in the Irish built environment, its buildings, monuments and even street names.[21]

Joyce's Dublin on film: Imperial political hegemony

Moving images of James Joyce's Dublin represent an invaluable historical record, equal to the photographic record of Charles Dickens's London. One of the main points of interest in early films of Dublin, a city with a population estimated at 350,343 inhabitants in the 1901 census, is indeed the possibility of seeing the streets of the city strolled by Leopold Bloom in *Ulysses*, one of the most influential literary expressions of urban modernity in the twentieth century.[22] In one case the connection between early film and *Ulysses* is particularly striking. In Mitchell and Kenyon's *Congregation Leaving Jesuit Church of St Francis Xavier, Dublin* (1902), the name of Father John Conmee can be seen on a poster outside the church, where he was Superior from 1898 to 1905. Revd. Conmee also appears in Ulysses, where he walks with Leopold from the presbytery to the Malahide Road. More broadly, it is important to see that early cinematic Dublin is as vibrant as it is in Joyce's novel. In his earlier work, *Dubliners* (1914), Joyce had portrayed Dublin as a peripheral town still traumatized by having ceased to be home to a parliament and the capital of a nation.[23] In *Dubliners*, as Jeop Leersen has argued, Joyce emphasized the stagnation and paralysis of a parochial and colonized place lacking the vitality and energy of true

21 Jeffery, "Introduction", 18.
22 James Joyce's Ulysses, which was first published after the First World War, chronicles the life of Leopold Bloom on 16 June 1904. James Joyce, *Ulysses* (London: Penguin Classics, 2000; first published in 1922).
23 James Joyce, *Dubliners* (London: Penguin Classics, 2000; first published in 1914).

modern urban culture and falling deeper into provincialism. By comparing the Dublin of *Dubliners* to that of *Ulysses* Leersen has observed how the same place can be see as peripheral at one moment and central at the next: in *Ulysses* Dublin is a vibrant and frenetic cityscape and is rendered by Joyce as a central twentieth-century metropolis.[24] On a visual level, the image of Dublin emerging from early actuality films is also determined by movement and fragmentation. Early films of Dublin did not present the formal and thematic complexity of Joyce's modernist narrative and yet cinematic Dublin appears a busy and central urban space where a sense of dynamism and movement is emphasized by the new medium.

In addition to *Congregation Leaving Jesuit Church of St Francis Xavier, Dublin*, Mitchell and Kenyon made other two films in Dublin. *Congregation Leaving St Mary's Pro-Cathedral, Dublin* (1901) was part of a show called "Living Dublin" and was shown at the Ancient Concert Rooms in Great Brunswick Street. This film consisted of a single shot taken in front of the church as the congregation leaves, and was part of film programmes designed to appeal to the Roman Catholic community of Dublin, members of which would recognize themselves on the screen. Arguably the most interesting film produced by Mitchell and Kenyon in Dublin, *Panorama of College Green, Dublin* (1902) represents the dynamic urban space of central Dublin and was filmed by operator and cameraman Louis de Clercq. In this film, College Green, a three-sided "square" in the centre of Dublin, is crowded with people pursuing their everyday activities, and a number of electric trams can be seen. College Green was the result of the establishment of the Wide Streets Commission, a government body aimed at reshaping the old city, and its project of widening Georgian Dublin's narrow streets.[25] Early cinematic Dublin is also a reflection of the Commission's work: the view of College Green and Foster's Place includes a number of landmarks such as King William III of England on horseback (bombed during the

24 Joep Leersen, *Remembrance and Imagination: Patterns in the Historical and Literary Representation of Ireland in the Nineteenth Century* (Cork and Notre Dame, IN: Cork University Press and University of Notre Dame Press, 1997), 227–228.

25 Joe Brady and Anngret Simms, *Dublin: Through Space and Time: c. 900–1900* (Ann Arbor: University of Michigan Press, 2002), 112.

annual Armistice Day Commemoration in 1928), the west side of the Irish Houses of Parliament (now Bank of Ireland), and Trinity College with its neoclassical statues. Similar scenes were filmed by other cameramen at the beginning of the twentieth century, and are now lost: *In and About Dublin* (Kineto, 1910), *Scenes of Irish Life at Dublin* (Vitagraph, 1913), *In and Around Dublin* (Heron, 1913), and *With the Bioscope Through Ireland* (Warwick Trading Company, 1900), which also included scenes filmed at College Green.[26]

In the work of scholars such as Vincent J. Cheng, Edna Duffy, and Tracey Teets Schwarze, the Dublin of James Joyce has been explored as the reflection of the views of a highly politicized author writing against imperial political hegemony.[27] Dublin is thus the capital of a nation that, as Schwarze puts it, "devises a nationalism that mimics the very structure of racism, ethnocentrism, and violence that Britannia perpetuated in order to subdue indigenous populations across the empire."[28] The exposure and derision of the consequent attempts to purify and "de-anglicize" Irish culture are at the heart of Joyce's writing. Early films of Dublin reflected the modern ease of movement in the city, but they are also deprived of a serious attempt to "de-anglicize" the city, and nationalist politics are clearly absent from representation.[29] The city is seen through the eyes, and the camera, of British filmmakers who often emphasized the submission of Dublin to London by filming events and celebrations which were expressions of British rule. For example, in the early 1900s a number of British

26 Another popular subject was the famous Dublin Horse Show at the Royal Dublin Society, which was filmed in 1900 by William Dickson for Biograph Syndicate, in 1906 by Charles Urban, and in 1909 by John Y. Brown for the London Cinematograph Company.

27 See: Vincent J. Cheng, *Joyce, Race, and Empire* (Cambridge: Cambridge University Press, 1995); Edna Duffy, "Disappearing Dublin: *Ulysses*, postcoloniality and the politics of space", in *Semicolonial Joyce*, eds Derek Attridge and Marjorie Howes (Cambridge: Cambridge University Press, 2000), 37–57; Tracey T. Schwarze, "Silencing Stephen: Colonial Pathologies in Victorian Dublin", in *Twentieth Century Literature*, vol. 43, no. 3 (1997): 243–263.

28 Schwarze, 243–244.

29 See Townshend, 39–43.

filmmakers gathered in Dublin to film the royal visits of Queen Victoria in 1900 and Edward VII in 1904. The visit of the Queen to Dublin was marked by two events: the procession through the streets of the city and the celebrations in Phoenix Park – the site of the murder in 1882 of Frederick Cavendish and Thomas Henry Burke, respectively the Chief Secretary for Ireland and the Permanent Undersecretary, carried out by members of the Irish National Invincibles. On 4 April Queen Victoria arrived at Kingstown and traveled to Dublin: her entry into the city was filmed by Birt Acres for the Northern Photographic Works (*The Queen's Entry into Dublin*). On that same day, the Warwick Trading Company produced *Her Majesty's Visit to Ireland*, a series including "Return of Earl Cadogan" (Henry Arthur Cadogan, Viscount Chelsea, a Conservative MP), "Royal Procession Passing Through Dublin", "21st Lancers Riding Through Dublin", and "Lancers, Guards and Volunteers". The procession was also filmed by Cecil Hepworth in *The Queen's Visit to Dublin*. A fragment of this film is extant and likely to be the only surviving film of royal visits to Ireland. "Late Queen's visit to Dublin: Royal Procession Entering the City Gates" shows the household cavalry passing through the city gates, which are opened for the mounted escorts and several open carriages and Queen Victoria's open landau. The streets are crowded with spectators waving their hats as the procession passes by. On 7 April, 52,000 children greeted the Queen at Phoenix Park: this event was filmed by William Dickson for Biograph Syndicate in a series called *Queen Victoria's Visit to Ireland* and including "Lord Lieutenant of Ireland Entering Phoenix Park", "The Queen Entering Phoenix Park", and "Presenting Bouquet to Queen in Phoenix Park." On this occasion, the site of nationalist murders was transformed in a territory subdue to the British monarchy and its empire. Four years later the visit of Edward VII and Queen Alexandra in Ireland was filmed by British Gaumont in a series called *The King's Visit to Ireland*, which included "Arrival and Procession, Dublin", "King and Queen at Belfast", and "Arrival of King and Queen at Kingstown and Dublin." Cecil Hepworth also produced a film of this event, *Royal Visit to Dublin*, which included "Arrival at Kingstown and Dublin", "Leaving Dublin", and "Royal Party at Belfast." An orderly image of the city emerged from these films placing Dublin solidly at the heart of the empire.

A trust in the advancement of urban life and the modern experience of the city was constantly reiterated in early films. For example, British Pathé produced a film of the opening of the Civic Exhibition of 1914 (*Dublin Civic Exhibition Opened by the Viceroy and Lady Aberdeen*), which was devoted to town planning, urban and rural life, industry and commerce and celebrated the advancements in these fields. However, British Pathé also produced a film of Dublin during the strikes of August 1913, when tram workers deserted their vehicles in protest against William Martin Murphy's Tramways Company, which forbade its employees to be members of the Irish Transport and General Workers Union. *Dublin Strikers*, which has survived, was part of a Pathé's Animated Gazette, which begun in 1910, and was the first British newsreel in the accepted sense.[30] The originality of its subject should be contextualized in the newsreel's informative nature: this kind of documentary film focused on a broader range of subjects than earlier factual films. There is only one earlier example of a film of strikes or riots in early twentieth-century Ireland. In 1907 Gaumont produced a film displaying a relatively unfamiliar and controversial theme in early factual cinematography. *Scenes of Belfast Riots*, which is now lost, included scenes of the strike by over 10,000 dock workers organized by trade unionist Jim Larkin. On 11 August, troops charged Catholic supporters of the Belfast Dockers' Strike on the Falls Road and opened fire, killing two workers, Charles Mullan and Margaret Lennon.[31] According to the distribution catalogues, this film contained images of the funerals of the victims. As we shall see in the following section, early actuality films of Belfast generally portrayed a rather different and less controversial city.

30 See Luke McKernan, *Topical Budget: The Great British News Film* (London: BFI Publishing, 1991), 1–5.

31 See John Gray, *City in Revolt: James Larkin & the Belfast Dock Strike of 1907* (Belfast: Blackstaff, 1985), 161–163.

The East Coast of Ireland: Wexford, Belfast and Derry

Not all towns and cities in Ireland shared the prosperity and the dynamism of Dublin. A dramatically different cityscape would have greeted a passenger boarding a Dublin, Wicklow and Wexford Railway Company train at Pearse Station on Westland Row on the south side of Dublin and getting off the train at Wexford. The arrival of such a train in Wexford was filmed by Mitchell and Kenyon in 1902 in an Irish version of the Lumières' *Arrival of a Train at La Ciotat*. In *Wexford Railway Station* the train arrives, its doors fly open and passengers disembark the train and walk towards the camera. In the same year, Mitchell and Kenyon also produced *Life in Wexford*, which included scenes of the Saturday fish and meat market at the Bull Ring with its traders, fishwives and members of the Royal Irish Constabulary. Other scenes were filmed from behind the grocers' stalls in the town centre and included a number of poorly dressed children staring into the lens. In this film Wexford, whose population was estimated at 11,168 inhabitants in the 1901 census, appears as a peripheral and provincial town. Hardly any trace of modernization emerges from these scenes, and *Life in Wexford* records an everyday life which had remained unchanged for decades.

With a population of 39,893 inhabitants, Derry was the second largest city in Ulster and, like Wexford, it was also filmed as a rural and poverty-stricken place in the only film of it which survives. *Derry Election Day*, now believed lost, was one of very few films taken in Derry and was filmed in 1913 by the Topical Film Company. Mitchell and Kenyon's *Cattle Market in Derry* (1902) thus remains the only identified moving picture of Derry before the Great War. Filmed at the end of March 1902, this film is composed of street and market scenes taken from a camera at street level. In contrast to the scenes of Dublin, *Cattle Market in Derry* displays a less modernized and a more rural Ireland. The streets of Derry are here being developed and construction work can be seen in the town centre. The cattle market held on Wednesdays on Rossville Street is the highlight of the film: the emphasis on scenes portraying farmers selling their cattle here reveals a more agricultural urban identity in contrast to the urban expansion of Dublin and Belfast.

Indeed, the only city which could compete with Dublin in terms of dynamism and modernization was Belfast, whose population was estimated at 349,180 inhabitants in the 1901 census. Indeed, economic prosperity and imperial themes were central to the image of Belfast in early films. As we have seen in the previous sections, Hepworth and Gaumont's films of the visit of King Edward VII and Queen Alexandra to Ireland both included scenes of official display taken in Belfast. First filmed by Alexandre Promio in 1897, Belfast was the subject of a number of actuality films made during the years preceding the Great War. By 1910 Belfast was one of the largest and most prosperous cities in Ireland.[32] Probably the film which best captures the prosperity of Belfast at the turn of the twentieth century is Mitchell and Kenyon's *Ride on a Tramcar through Belfast*. Filmed by Ralph Pringle's North American Animated Photo Company in May 1901 and exhibited at the Ulster Hall during the same month, this film presents the city from the front of a horse tram as it runs through the spacious avenues of central Belfast.[33] *Ride on a Tramcar Through Belfast* offers a glimpse of the city's horse tramway during one of its last years in service. In this film, which begins with a shot of Bedford Street, the tram turns past Castle Junction and Donegal Place onto Royal Avenue and continues along the High Street, where the base of the Albert Memorial Clock, one of Belfast's most prominent landmarks, can clearly be seen. These streets are crowded with men, women and children, dressed both in smart and working clothing, pursuing their everyday activities or staring at the tram and the cameraman as they pass by. A number of bicycles, horse-drawn trams full of passengers and advertising boards for John Arnott, Boots, and Hudson Soap can be seen and contribute to a portrait of a rapidly moving cityscape.

Other street scenes were filmed during those years by travel cameramen for several British companies. Warwick Trading Company's *With the Bioscope through Ireland* included the now lost "Royal Avenue, Belfast" and "High Street, Belfast." *Belfast Street Scenes* (1898 ca., unidentified) has survived and, filmed with a camera at street level, it shows the same places later

32 Foster, 439–440.

33 In 1905 these trams would be replaced by the new electric tram system.

filmed by Ralph Pringle for Mitchell and Kenyon. In this film, a number of views of Belfast start with a shot looking down the High Street to the Albert Memorial Clock Tower, showing a horse-drawn tram and pedestrians. The view then cuts to Queen's Bridge, which had also been filmed by Alexandre Promio, and which is followed by views of the unloading of a ship in the dockyards.

Belfast was also the subject of a few topical films made during the early 1910s. *Review of Ulster Division at Malone Park by Sir H. Calmont* (1914, unidentified) was shot at Malone Park and in an unidentified street of Belfast. Military columns including soldiers with bicycles, carts and wagons march through the park and the street, while a number of spectators watch the parade. A close-up of the Union Flag separates the scene at the park and that of the street. The definitive icon of British imperialism, the Union Flag appeared waving on buildings and flagstaff in a number of early films: in 1901 Cecil Hepworth even filmed *Flag of Britain*, which exclusively represented the Union Flag waving on its flagstaff. Another film revealing unionist views is *Sir Edward Carson and Lord Londonderry Outside the Committee Rooms in Belfast* (1912, Gaumont), which includes a view of the Ulster Hall, a unionist or loyalist urban sight, and thus defines Belfast, which had a larger protestant and unionist population than Dublin, as yet another orderly reflection of the imperial system.

The third city of Ireland: Street life and the International Exhibition in Cork

A strongly Irish nationalistic city, Cork grew rapidly during the 1840s, when people migrated from the countryside because of the famine, and experienced a number of industrial and municipal improvements. For example, the railway reached the city in 1849. The population later declined because of immigration to Canada and the United States, but at the turn of the twentieth century, with a population of 76,122 inhabitants, Cork was one of the most prosperous cities in Ireland.[34] The city is likely to have been filmed for the first time by the Warwick Trading Company cameramen who made *With the Bioscope Through Ireland* and *The Bioscope in Ireland* in the early 1900s. Both series include scenes of Patrick Street, and the latter also includes "Panorama of the Exhibition Grounds, Cork." Years later, Cork was also the subject of a Vitagraph film: *Cork and Vicinity*, which was made in 1913. Unfortunately, none of these films has survived.

The only early films of Cork available today are those produced by Mitchell and Kenyon. *Tram Ride from King Street to Patrick's Bridge, Cork*, was filmed in the city centre in 1902 from the front of an electric tram, a technology introduced in Cork in 1898. The tram leaves from King Street (now McCurtain Street), stops at the junction with Coburg Street, turns left into Bridge Street, and continues through Patrick's Bridge over the River Lee to Patrick Street. The streets of Cork had been regenerated in the nineteenth century and here appear as busy, spacious, and apparently prosperous. Mitchell and Kenyon present a view of Cork as a settled part of the British Empire, and this is emphasized by the presence of a Union Flag waving from a building in Patrick Street. In the same year, Mitchell and Kenyon also filmed *Views of the Grand Parade, Cork*, which included two sequences. The first resembles the film *Tram Ride from King Street to Patrick's Bridge, Cork* but presents the tram's journey in the opposite direction. In the second sequence, the camera is set up on the left of the Grand Parade. Both scenes record a very busy urban space with people engaged in shopping, delivering goods, travelling by horse and electric tram (revealing the interaction between old and new forms of transport), as well as a steam-roller, and street-sweepers. Electric street lamps, which had been introduced in 1898, are also filmed.

Mitchell and Kenyon's *Albert Quay in Cork* (1902) is a long panning shot of the activities on the quayside on a Sunday morning: horse traffic, fishermen on their boats, and local people on the way to church. In the background, we can see the area of Montenotte, which had become home to the wealthier merchants and the middle classes during the second half

34 See Foster, 210–212.

of the nineteenth century.[35] Their vacated houses in the inner city became homes for the working classes and the poor, often living in dreadful conditions: the slums, as it is to be expected, are not shown. On the contrary, scenes filmed by Mitchell and Kenyon outside the city churches project an image of prosperity of a city rooted in the British Empire. *Congregation Leaving St Patrick's Church in Cork* and *Congregation Leaving St Mary's Dominican Church in Cork* (1902), for example, portray the smartly-dressed residents of the suburbs of Montenotte, Tivoli, and Blackrock.

In these films, Cork appears as a settled part of the British Empire: no sign of political nationalist activism can be detected, the city appears wealthy and dynamic, and the familiar sight of the Union Flag contributes to defining the city streets as part of the Union. Arguably, the most prosperous image of Cork in early films is offered by Mitchell and Kenyon's films of the Cork Industrial International Exhibition (1902–1903). The exhibition was proposed by the Lord Mayor of Cork, Edward Fitzgerald, and took place in an area of parkland between the Cork County Cricket Grounds and Wellington Bridge (now Thomas Davis Bridge). By all accounts, it was a spectacular exhibition, including a great number of halls displaying industrial, agricultural and artistic wares, pavilions, kiosks, walks, tea houses, a water chute and a switchback railway, and attracting visitors from all over Europe. Edward VII and Queen Alexandra visited the exhibition during their trip to Ireland in 1903. Mitchell and Kenyon produced six films recording it in different stages. *Preparation of the Cork Exhibition Grounds and Erection of Buildings* was filmed one month before the opening, recording construction in progress, and including a panning view of Cork's west suburbs near the park. *Panorama of Cork Exhibition Grounds* consists of a number of panning shots recording purpose-built exhibition buildings, themed halls celebrating aspects of Edwardian progress, an industrial hall, a machinery hall, and the concert hall. In *Trade Procession at Opening of Cork Exhibition*, the camera was fixed on the ground on the Western Road and records the opening ceremony on Mayday. In this film we can see a procession of local tradesmen and the associate Guilds of Cork,

35 See John Crowley, *Atlas of Cork City* (Cork: Cork University Press, 2005), 9–10.

including Irish National Foresters members in full regalia, trade unions members, Friendly Societies and local church guilds, the Irish Language Society holding a banner in Gaelic and a tableau vivant of Adam and Eve, and the Lord Mayors of Dublin, Derry, Galway, Limerick, and Waterford. As the procession passes by, a large crowd waves and cheers. *Arrival of VIPs for Official Opening of Cork Exhibition*, *Lord Mayor of Cork arriving for Official Opening of Cork Exhibition*, and *The Visit of the Duke of Connaught C-I-C Forces in Ireland* and *Prince Henry of Prussia to Cork Exhibition* all record the arrival of dignitaries, the Royal Constabulary and the Lord Mayor of Cork.

As with world's fairs in British cities, exhibitions held in other territories of the empire served an ideological purpose and displayed either cultural otherness or, as in the case of the International Exhibition in Cork, the technological achievements of the empire. Films such as those produced by Mitchell and Kenyon record this event and emphasize both its local dimension, with the presence of local dignitaries, guilds, unions and societies, and its international appeal, with a number of foreign dignitaries and an opulent display of industrial, agricultural and artistic wares. Cork, as much as Dublin and Belfast, was filmed as a part of the Union with a distinctive national culture. Here Ireland's subjugation to Britain was made to appear unproblematic by British filmmakers. British films of Ireland do not reveal the country as Joyce's "old sow that eats her farrow", but as a modern and at times picturesque territory of the empire and the Union with Britain.[36]

Cinematic cities are defined by their omissions just as by their inclusions. Similarly, the modern city is defined by what it keeps out, by the exclusion of its opposite. As Timothy Mitchell suggests, the Western city determines itself as the place of order, cleanliness and power by representing the "outside" – the Eastern city – as disordered, dirty and unruly.[37]

36 Stephen Dedalus describes Ireland as "the old sow that eats her farrow" in James Joyce's *Portrait of the Artist as a Young Men* (London: Penguin Classics, 2000; first published in 1916), 216.

37 Timothy Mitchell, *Colonising Egypt* (Berkeley: University of California Press, 1992), 65.

Early cinematic representation of British and Eastern cities contributed to the process of establishing one's identity through a visual medium. In the following chapters, I investigate this process as it emerges from films made by British filmmakers in cities in colonized territories.

PART III

The Empire

CHAPTER 6

Travel Cinematography in India

Where the author meditates on the political stakes emerging from films of Varanasi, Delhi and Colombo, and reflects on the definitions of "home" and "journey".

Opened in the same year as the Crystal Palace in 1851, James Wyld's Globe in Leicester Square symbolized the growing interest in geographical spectacles in Victorian Britain. It consisted of a globe sixty feet in diameter with the surface of the Earth shown inside the building and viewed by the public from a series of galleries and iron steps.[1] Wyld's Globe was one of many exhibits in this era which encouraged a new image of the globe resting on the idea of its visibility. As Denis Cosgrove suggests, this kind of image of the globe emerged from a series of non-linguistic ways of seeing and representing space as an "icon for the interrelated processes of connection, communication, and control which characterize modernity."[2]

Indeed, as we have seen in previous chapters, the Victorian era saw a great number of colonial exhibitions, world's fairs and technological displays, and pioneer filmmakers filmed places where colonial and exotic displays were exhibited, such as the Crystal Palace, the London Zoo and the Glasgow Exhibition. These sites, together with panoramas, dioramas, cosmoramas and theatres, reflected a new world-view that transformed exotic landscapes in records of cultural practice. Films of exhibitions, as

1 See Liza Picard, *Victorian London: the Life of a City 1840–1870* (London: Weidenfeld & Nicolson, 2006), 257.

2 Denis Cosgrove, *Apollo's Eye: a Cartographic Genealogy of the Earth in the Western Imagination* (Baltimore and London: The John Hopkins University Press, 2001), 206–207.

seen in the previous chapters, offered a glimpse of the informative ethnographical possibilities of the medium. Travel films also reflected life in exotic territories in a way which was new and yet drew upon pre-existing expressions of cultural practice. At the turn of the century, pioneer filmmakers continued a tradition of exploration and representation which, from the end of the eighteenth century, had seen European travellers attempting to record the experiences of geographical exploration and to reproduce them in different ways for their fellow countrymen.

Cinema played a significant role in imperial visual culture as it showed its potential as a tool to record evidence from distant destinations. Colonial landscapes and architecture had already become familiar to audiences at fairs and exhibitions, and filmmakers presented a new world-view to British audiences whose geographical knowledge had grown in the course of the nineteenth century. Filmmakers perpetuated the tradition of empirical observation and amateur scientific investigation which had previously been embodied in travel photography and writing, and which was concerned with the imperial project of world knowledge and its political and cultural boundaries. As Michael Chanan has argued, pioneer cameramen brought back to Britain images to be added to the catalogue of the exotic, and displayed their instinct to capture physical reality in the manner of the Victorian explorer, the idealized man-of-action of the nineteenth century.[3] Subsequently, British cameramen travelled all over the world filming exotic landscapes and their inhabitants, as well as the remains of ancient civilizations. The exoticism of the territories filmed was a function of their physical distance from British audiences, and an attempt to locate something lacking within the self in a strange and foreign landscape. The exotic was thus a landscape outside immediate reach, but made concrete in filming.

Most films represented countries subjugated, directly or indirectly, to the British Empire, its colonies and its dominions. In this and the following chapters, I investigate the ways in which actuality films portrayed cities in colonized territories and other exotic terrains, and explore how the interest for geographical spectacle extended to early films of urban spaces in colonized territories.

3 Michael Chanan, *The Dream that Kicks: the Prehistory and Early Years of Cinema in Britain* (London and New York: Routledge, 1996; original edn 1980), 238–239.

Culture of travel and exploration: Towards a cinematic encounter

Travel films of cities caught the attention of the press from their early days. In the summer of 1896, Henry William Short photographed a series of fourteen films in Spain and Portugal for the company owned by R.W. Paul. That October, *The Era* commented:

> The latest series, entitled "A Tour in Spain and Portugal", is not only entertaining but highly instructive. Street scenes in Madrid, Cadiz, Lisbon, and Seville are happily depicted. We see worshippers leaving the church of San Salvador, Seville; and other excellent illustrations of Peninsular life include a fish market scene in Lisbon; procession of bull-fighters at Seville; the bull-fight, an exciting picture; and a cave on the coast of Galicia, one of the most beautiful realisations of the sea that we have ever witnessed.[4]

Three weeks earlier, *The Era* recorded Paul's plan to arrange a tour throughout Egypt, Persia, and Japan, for the purpose of making new films.[5] As *The Era* records, from the very first days of the cinematograph, companies such as Paul's and the Warwick Trading Company sent their cameramen all over the world, from Stockholm and Constantinople to New York and the Swiss Alps, to make new films to be shown in Britain. Thus British actuality films were affected by what Giuliana Bruno, in her study of early cinema, has called "a real travel bug."[6] Travel films of the distant countries and their cities were distributed and exhibited both in the United Kingdom and in the colonies. Paul's films, for example, were exhibited at the Empire Theatre of Varieties in Johannesburg as early as May 1896. These films were among the most popular types of films in the early days of cinematography. Rachael Low suggests that British pioneer filmmakers looked at travel scenes as the fulfilment of the proper function of the cinematograph:

4 *The Era* (London, England), 31 October 1896, 19.

5 *The Era* (London, England), 10 October 1896, 19.

6 Bruno, *Atlas of Emotions, Journeys in Art, Architecture, and Film*, 76.

> The enlarged moving picture gave these audiences for the first time in history a direct reproduction of the surface movement of life in other countries, and so a more vivid impression of the world than could ever be gained by still photographs or drawings of the contemporary popular periodicals.[7]

Travel films, as Low suggests, emerged as a record of exotic foreign countries, and their emphasis fell consistently on Asia, the Orient, the South Seas, and Africa.[8] British pioneer filmmakers thus created series of views and scenes of exotic places just as photographers had already done in the nineteenth century so that, together with travel photography, exhibitions, theatre, travel writing and painting, early cinematography contributed to the imperial collection of knowledge, or to what Thomas Richards has defined as "a fantasy of knowledge collected and united in the service of state and Empire" and which aimed at a superintending unity of knowledge of the world.[9]

Following this stream of thought, Michel Foucault argued that the political stakes of visual culture emerge from all its forms, including those that claim supposed neutrality. Foucault investigated the hierarchical classification systems in western natural history, arguing that museums and exhibitions construct cultural master narratives by imposing a system of beliefs as the dominant cultural manifestation of a given era.[10] Victorian and Edwardian visual culture put forward the imperial theme through exhibitions, photography and museums. When moving pictures came into being, their makers drew upon subjects and themes already exploited by earlier forms of visuality and sustained the imperial narrative as the chief

7 Rachel Low, *The History of the British Film: 1896–1906* (London: Gerges Allen & Unwin, 1948), 71–74.

8 Rachel Low, *The History of the British Film: 1906–1914* (London: Gerges Allen & Unwin, 1949), 152–153.

9 Thomas Richards, *The Imperial Archive: Knowledge and the Fantasy of Empire* (London and New York: Verso, 1993), 4–6. See also Thomas Richards, *The Commodity Culture of Victorian England* (Chicago and London: Stanford University Press, 1990), 17–72.

10 See Michel Foucault, *The Order of Things: an Archaeology of Human Sciences* (New York: Vintage, 1994), and Michel Foucault, *Power/Knowledge: Selected Interviews and Writings, 1972–1977* (New York: Pantheon, 1981).

manifestation of their era. Before the birth of cinematography, British theatre and the entertainment industry helped constitute a modern colonial imaginary, a Western image of the East, by disseminating visual information focused on both the object and its surrounding space. Exoticism as a process by which space is dramatized, according to Edward Ziter, emerged from both on and off-stage narratives, and helped constitute a recreation of self and other in the consciousness of the viewer.[11] This exotic reinvention of space then migrated to representational strategies of early films dramatizing distant places within the familiar space of the music hall.

The apogee of colonial domination, as Ella Shohat has argued, coincided with the beginnings of cinema, which fostered imagined communities and made it possible to record glimpses of the margins of the empire previously captured only through travel, photography and tourism.[12] Representational discourses of early cinema were bound to an imperial archive of knowledge which was both scientific and popular, with world's fairs displaying reproductions of exotic streets scenes, with human subjects, called "ethnographic villages" for a vast popular audience.[13] These displays emerged from the same complex network of visual popular representations which also influenced early cinema's reproductions of vast exotic cities and movement within these urban spaces. Central to the representational strategies of early films was imperialism, although its significance in visual culture has been much less thoroughly investigated in relation to films than it has in relation to painting and other arts.[14]

11 Edward Ziter, *The Orient on the Victorian Stage* (Cambridge and New York: Cambridge University Press, 2003), 3.

12 Ella Shohat, *Unthinking Eurocentrism: Multiculturalism and the Media* (London and New York: Routledge, 1994), 100–109.

13 On ethnographic villages, see Alison Griffiths, *Wondrous Difference: Cinema, Anthropology and Turn-of-the-century Visual Culture* (New York: Columbia University Press, 2002), 46–85.

14 The idea that the concept of empire belongs to the centre of British visual culture has been put forward, for example, in Tim Barringer, Geoff Quilley and Douglas Fordham, *Art and the British Empire* (Manchester and New York: Manchester University Press, 2007).

W.J.T. Mitchell has argued that under imperialism landscape becomes a medium of cultural expression and a symbolic form in its own right. Eighteenth-century English landscape painting, according to Mitchell, was characterized by a double movement to the foreign fields towards which empire directed itself and by a renewed interest in the nature of the home landscape.[15] At the end of the following century, non-fiction films manifested this two-directional movement by depicting the imperial centre, as we have seen in the previous two chapters, and by recording life in the distant territories of the empire. Addressing colonialism as a process of imagining, Nicholas Thomas has argued that nineteenth-century travel was a peculiarly modern activity insofar as it entailed an attitude of extension and displacement away from traditional ties. The modern traveller, Thomas has argued, was deconstructed and reconstructed through activities as diverse as "collecting Egyptian mummies and drawing crusader castles."[16] Following this train of thought, Felix Driver has argued that, in the nineteenth century, exploration was a heterogeneous activity conceived of as a particular kind of travel associated with the sight of new places and people. Geographical knowledge was produced in a variety of contexts, both covering a scientific purpose and directly sustaining the imperial and colonial project: mapping the earth through different ways of collecting evidence was thus aimed at both knowing and subduing the world.[17] Driver, in a chapter written with Luciana Martins, has demonstrated how the mapping of the world was central to the project of the Enlightenment because it involved translation through inscription and taking the measure of the world, observing its phenomena and making its data accessible to others. In the name of the Enlightenment, navigators like Thomas Cook

15 William J.T. Mitchell, "Imperial Landscape", in *Landscape and Power*, ed. William J.T. Mitchell (Chicago and London: University of Chicago Press, 2002; first published in 1994), 13–14.

16 Nicholas Thomas, *Colonialism's Culture: Anthropology, Travel and Government* (Oxford and Cambridge: Polity Press, 1994), 5, 55.

17 Felix Driver, "Geographical Knowledge, Exploration and Empire", in *Geography Militant: Cultures of Exploration and Empire*, ed. Felix Driver (Oxford and Malden, MA: Blackwell, 2001), 21.

and Jean François de Galaup, and terrestrial explorers like Alexander von Humboldt and Mungo Park sought to make the world an orderly place. Such people, Driver and Martins continue, "imagined the establishment of vast archives of texts, images, artefacts, and specimens, patiently assembled, through which geography and natural history of the earth could be made known"; their project involved an engagement with particular places and sites and with the production of visual images in situ and the making of views on the spot.[18] By producing visual images of the colonies, early films also contributed to observing the world and making its data accessible to the Western viewer. Films thus mediated what Gyan Prakash has defined as a dual process by which colonial rulers sought to constitute the "native" as their inverse image while the "native" was pressured to identify with the colonizer.[19] The strategies enacted by Western colonizers to define the colonized as the other found in early cinema a new form of representation.

Inderpal Grewal has discussed the terms travel and home in similar terms, arguing that travel is an ontological discourse central to the relationship between self and other and, on the other hand, that home is a crucial category within European travel because it is the space of return and of consolidation of the self enabled by the encounter with the other.[20] As we look at early British cinema, home was both a figurative place of return – the familiar, Western colonizing identity – and literally a physical place of return, the British countryside, the city and ultimately the music hall or, later in the 1900s, the cinema palace, where the exhibition of what was recorded by the travelling cameraman enables the Western self to consolidate its identity.

18 Felix Driver, and Luciana Martins, "Views and Visions of the Tropical World", in *Tropical Visions in an Age of Empire*, eds Felix Driver and Luciana Martins (Chicago and London: Chicago University Press, 2005), 9.

19 Gyan Prakash, *After Colonialism: Imperial Histories and Postcolonial Displacements* (Princeton, NJ: Princeton University Press, 1995), 3–6.

20 Inderpal Grewal, *Home and Harem: Nation, Gender, Empire, and Cultures of Travel* (Durham and London: Duke University Press, 1996), 3–8.

From travel photography to travel cinematography: Geographical knowledge and the British Empire

Before the cinematograph, geographical knowledge found a form of expression in travel photography, which had been a useful instrument for ethnographic surveys since the mid nineteenth century and succeeded in stimulating the imperial imagination. The conventional visualization of the Eastern city, which focused on folkloristic and monumental urban spaces and activities, had been established by travel photography and illustrated magazines containing woodcut prints. Among these, *The Illustrated London News* often published images and scenes of non-European cities. In January 1903, for example, a series of woodcut prints representing the imperial city of Fez, Morocco, included a view of the city, a street scene, a gun maker's shop, the interior of a fondak (merchant's court) and the gates of the city.[21] These prints are remarkable for their representation of everyday life as an exotic and distant reality. Similarly, photographers such as John Murray, Samuel Bourne and Josiah Rowe took a number of photographs of Indian cities, often published in *The British Journal of Photography*, and their work included panoramas of Bombay, Calcutta and Bundi.[22] Their work also helped to make Western viewers familiar with landscapes which, in most cases, they would never have seen in person.

British photographers were concerned with the scenery and the monuments of distant places and the customs of their inhabitants. As John M. MacKenzie puts it, on one hand, photography embodied the progress of technical and scientific discovery of the period and, on the other hand, it "literally captured the world on glass and paper."[23] Photographers set out for Africa, India, Burma and China, and, satisfying a new taste for the exotic by providing the West with an image of the world remod-

21 *The Illustrated London News*, No. 3324, Vol. CXXII (Saturday, January 3, 1903): 4.

22 See Vidya Dehejia, *India through the Lens: Photography 1840–1911* (Washington: Freer Gallery of Art and Arthur M. Sackler Gallery, 2001).

23 MacKenzie, *Propaganda and Empire*, 21.

elled through economic and political control. The birth of photography influenced the history of geographical visual knowledge of Britain's Age of Empire. Photographic images and aesthetics of the colonized world formed a collective colonial archive aimed at expressing and articulating ideologies of imperialism. A large part of the Victorian imagination was built through photographic images related to geographical knowledge and imperial discourse. As James R. Ryan has explained, in its ability to conquer space and time photography was a powerful means of classification and visualization of the non-European world, and helped Victorians to familiarize themselves with exotic views, enabling them to travel through, explore and even possess those spaces. Ryan suggests that photographs were aimed at consumption within middle- and upper-class circles in Britain, and were revealing of both the imaginative landscapes of imperial culture and the physical spaces or people represented.[24] By enabling viewers to visualize reality from their doorstep to the world beyond, photographs helped them to *picture place* both literally and figuratively, and to know the world and situate themselves in space and time. Ryan and Joanna Schwartz have asserted that photographs participated actively in the formation of geographical knowledge and "were ideally suited to empiricism and the nineteenth century passion for collecting, classifying and controlling facts, whether in the pursuit of comprehensive knowledge or the conduct of imperial administration."[25] Early films of non-Western spaces and people also helped conquering colonized spaces through the visualization and the domestication of a potentially hostile landscape. Just as with photographs, through films the viewers could *picture place* and, by subduing a potentially threatening sight, situate themselves in the empire.

The themes and subjects displayed in travel photography are strikingly similar to those of early travel films. Hitherto, the work of pioneer

24 James R. Ryan, *Picturing Empire: Photography and the Visualization of the British Empire* (London: Reaktion Books, 1997), 20–35, 214–219.

25 Joanna M. Schwartz and James R. Ryan, "Introduction: Picturing Place: Photography and the Geographical Imagination", in *Picturing Place: Photography and the Geographical Imagination*, eds Joanna M. Schwartz and James R. Ryan (London and New York: I.B. Tauris, 2003), 1.

filmmakers has never been investigated in depth as a part of the imperial project which aimed to collect objects and visual representations of the colonized world in the attempt to classify knowledge. As James Clifford has argued, since the turn of the century, objects collected from non-Western sources have been classified either as cultural artefacts or as works of art, and mass-produced commodities have been less systematically valued. This classification resulted in mass-produced visual representations of non-Western places and people, such as films and illustrated magazines, being less valued. Clifford has also suggested that object systems of art and anthropology are not immutable, that they have changed and they continue to change.[26] The neglected films of British pioneer filmmakers now deserve to be regarded as worthwhile sources in the process of investigating the ways in which the colonizers looked at the colonized world. The understanding of the ways in which modern activities such as travel and photography were perceived during the nineteenth century is thus essential to place early films within a framework which emphasizes their contribution to the development of a colonial imaginary.

Though similar in themes and subjects, a significant difference exists between the consumption of photographs and films: the cinematograph made images of colonial urban spaces available to a much larger audience than that for travel photography. As John MacKenzie has demonstrated, illustrated magazines, photographic books as well as the stereoscope were relatively expensive and were aimed at upper middle-class and aristocratic citizens.[27] It was only with the cinematograph that photographic techniques became widely available to all classes. Early cinematography, whose audience was primarily working-class and lower middle-class, brought images of exotic urban landscapes to people who had hitherto been largely unacquainted with such scenes and views. As MacKenzie argues, picture postcards had also reached the height of their popularity at the turn of the twentieth century and contributed to the democratization of the visual

26 J. Clifford, *The Predicament of Culture*, Cambridge, MA, and London: Harvard University Press, 1988, pp. 223–229.
27 MacKenzie, *Propaganda and Empire*, 20–21.

image.[28] However, panoramic films and travelogues reached a larger audience, while offering a more realistic impression of the world than earlier media, as these images were photographic and moving. Their realism impressed their original audiences as a novelty just as now in the twenty-first century viewers see in them impressive glimpses of a long-gone time.

The exoticism of everyday life: Early films of cityscapes in colonized societies

The exotic has long been a prominent category of difference in Western culture, and the term "exoticism", as Roy Porter and George Rousseau suggest, has been used to define everything which defied normalcy and transgressed the norms of Christian, civilized, and rational Europe.[29] During the nineteenth century European fantasies of the exotic grew massively, and despite the development of revival styles in art like Victorian Gothic architecture and Pre-Raphaelite painting and poetry, the fascination with the exotic in popular culture was increasingly identified with other places and other peoples rather than other times. For example, from the mid 1780s exotic people and events became one of the main subjects of topical shows in London: John O'Keefe's pantomime *Omai, or a Trip Round the World* (1785) and *Tippoo Saib or British Valour in India* (1791) were only two of the most successful plays offering fantastic visions of the East. From the mid-nineteenth century trips around the world and images of exotic native life were the subject of exploratory and didactic panoramas exhibited in Britain for large urban audiences.[30]

28 MacKenzie, *Propaganda and Empire*, 21–22.

29 George S. Rousseau and Roy Porter, "Introduction", in *Exoticism in the Enlightenment*, eds George S. Rousseau and Roy Porter (New York and Manchester: New York University Press, 1990), 46–65.

30 MacKenzie, "Empire and Metropolitan Cultures", 275–280.

According to Homi K. Bhabha, the Western colonial impulse in the eighteenth and nineteenth centuries was supported by the conservative Western representations of foreign cultures.[31] The politicization of point of view in the means of representation should therefore be central to the study of early films of colonial cities and is evident in the ways in which filmmakers omitted, emphasized and selected their orientalist subjects. Edward Said defined Orientalism as a discourse determined by the ways in which European culture succeeded in managing and producing, dominating and restructuring an Orient that meant more than the sum of empirical knowledge about it. The politics and the military power, the ideology, science and imagination of the Orient were, according to Said, undermined by a Western authority over the colonized countries which aimed at understanding, controlling, manipulating and even incorporating a manifestly different world. For Said the Orient was an idea or a creation consistent with Orientalism and dependent on a series of relationships between Westerner and the Orient, where the latter is subordinate to the former.[32] Said's work suggests that early film audiences would have been unable to separate moving images of India or South Africa from the status of those places in their minds as colonized and subdued territories. Nevertheless, Said's implication that discourses of cultural difference always involve hostility is not the only idea applicable to the study of visual representations of the Orient in Victorian and Edwardian times. Nicholas Thomas, for example, has questioned the presumption that colonial discourses exclusively depict colonized people pejoratively, and put forward examples such as the appreciative images of Indian individuals and landscapes made by William Hodges, the Daniells, Henry Salt and James Fraser in the early nineteenth century.[33] In their popular aquatints, India is not a place, but a series of views and scenes made available for quotation and imitation. As I shall explain in the following section, a similar argument can be put

31 Homi K. Bhabha, *The Location of Culture* (London and New York: Routledge, 1994), 66–70.

32 Said, *Orientalism: Western Conceptions of the Orient.*

33 Thomas, 20–27, 54–55.

forward in relation to early actuality films of Indian cities and the ways in which they renewed Orientalist representational processes.[34] This approach should take into account the ways in which the British Empire and its cultural agencies aimed both at constructing affinities and at creating otherness within the imperialist system.

The definition itself of a colonial city requires an explanation. In his discussion of the subject, Anthony D. King has suggested a more-than-semantic adjustment from "colonial city" to "city in a colonized society or territory", as this definition turns our attention to the society, the territory and location, where it is, the process of colonization and the city which results.[35] In this city, King argues, the power is in the hands of a non-indigenous minority superior in terms of military, technological, economic and social organization, and subduing an ethnically, culturally and religiously different majority.[36] The cinematic image of the "city in a colonized society or territory" was produced by Western filmmakers for Western viewers, and it acquired coherence not because of its individual familiarity, but by conforming to the stereotypical construction produced by Western observers.

Hundreds of films of cities in colonized territories were filmed in the period 1895–1914. The themes emerging from these films can be classified into two broad groups reflecting the distinction between settler colonies, or settler societies, and colonies of occupation, or subject colonies. The primary difference between the two is that settlers tended to stay permanently in settler colonies while, in colonies of occupation, the "settlers"

34 Orientalism has recently been reassessed within the broad heterogeneous field of visual culture, museology, geography, history and archaeology. Early films of colonial cities should also be included in this group as their study contributes to the understanding of the heterogeneity of Orientalism and popular visual culture. See, for example, Jocelyn Hackforth-Jones and Mary Roberts (eds), *Edges of Empire*: Orientalism and Visual Culture (Malden, MA, and Oxford: Blackwell, 2005).

35 Anthony D. King, "Colonial Cities: Global Pivots of Change", in *Colonial Cities: Essays on Urbanism in a Colonial Context*, eds Robert Ross and George J. Telkamp (Dordrecht, Netherlands: Martinus Nijhoff, 1985), 8.

36 King, 9–10.

consisted of a small but powerful group of white planters managing the exploitation of resources and safeguarding the interests of the metropolitan state.[37] On the one hand, the image of the city which emerged from actuality films made in colonies of occupation, such as India, privileged the representation of exotic local life. Pioneer filmmakers sought to record the exoticism of the "native" people and quarters and created a series of often picturesque views of otherness in the form of colonized individuals, outmoded everyday activities and cityscapes. Early travel cinematography also recorded the military and political influence of British rule on the colonized territories by filming ceremonials such as the Delhi Durbars and "white" districts in the cities subdued to the Empire. On the other hand, films of settler colonies exclusively offered a representation of urban space which emphasized the non-indigenous presence in those territories and the power of the ruling class. A relatively small number of films were made in settler colonies, fewer represented urban spaces, and even fewer have survived. Early films of Canada such as *Farming in N.W. Canada* (Urban, 1910), *The Resources of Canada* (Urban, 1914) and *The Wheat Fields of Canada* (Warwick Trading Company, 1908) focused on the popular perception of Canada as the granary of Britain, an inviting settlement to move to. A few films of Canadian towns were also produced. Montreal was filmed as early as in 1895 by Birt Acres, and Winnipeg and Montreal were filmed in 1901 by Joe Rosenthal for the Warwick Trading Company. Both Winnipeg and Montreal developed as European settlements and, although these films are now thought to be lost, it is safe to argue that their cinematic representation focused on non-indigenous life in Canada: imperial sentiments manifested themselves in a number of ways in Canada – as in the filming of tributes to Queen Victoria on her death in 1901 – and its towns were ideal places of identification for British audiences.[38]

37 See Bill Ashcroft, Gareth Griffins, and Helen Tiffin (eds), *Post-Colonial Studies: The Key Concepts* (London: Routledge, 2003), 93.

38 See John H. Thompson, "Canada and the 'Third British Empire', 1901–1939", in *Canada and the British Empire*, ed. Phillip Buckner (Oxford: Oxford University Press, 2008), 87–106.

While no film of European settlements in New Zealand is recorded, we know that in 1901 Joe Rosenthal filmed the streets of Sydney, Australia. The same city is also the subject of *Sydney Street Scenes* (1912), made by cameraman *Emile Lauste* for the Empire Company. *Capital Cities of Australia* was a film produced by Davison's Film Sale Agency in 1914. These films are not extant but a film of Bourke Street, Melbourne, filmed for Imperium Film in 1911 has survived and it shows a large and crowded street which is reminiscent of British urban thoroughfares. The street in that film is reminiscent of a print of Collins Street in Melbourne which had been published in *The Illustrated London News* in 1890, and which displays a white, rich city, where no "natives" were to be seen.[39] Cinema audiences in Britain would have been struck, not by the exoticism of these territories, but by the ways in which the identity of a city like Melbourne, which Åsa Briggs defined as a Victorian community overseas, resembled that of their own cities.[40]

These few exceptions are all the recorded British films of Australian and Canadian cities made between 1895 and 1914. In what follows I investigate the representation of urban space in colonies of occupation by focusing on the exotic, on one hand, and the British presence, on the other. I will start with films of Indian cities, in particular Varanasi and Delhi. In the following chapter I will venture outside the British Empire and discuss the cinematic representation of East Asian cities, paying particular attention to urban environments created to different degrees by European invasion, such as Hong Kong, Shanghai and Singapore. The final chapter is devoted to actuality films of South African cities, against the background of the Boer War, and to the representation of urban spaces in the Islamic East, with particular attention to Egypt. Most of the films I refer to are now thought to be lost, and more in-depth attention is given to the analysis of the fewer surviving films. Nevertheless, I also take into account non-extant films both because of their statistical relevance and their often revealing titles.

39 *The Illustrated London News*, No. 2646, Vol. XCVII (Saturday, January 4, 1890): 16–17.

40 See Åsa Briggs, *Victorian Cities* (London: Penguin, 1990).

A passage to India: Imperial narratives and cinematic exoticism in South Asia

Prem Chowdhry has applied the definition "empire-cinema" to British and Hollywood films made during the 1930s and 1940s, and which emerged as powerful propaganda vehicles and "projected a certain vision of the empire in relation to its subjects."[41] A number of films with India as their central theme, Chowdhry suggests, reached the height of their popularity in the 1930s and promoted a vision of Britain which encouraged British working-class audiences to identify themselves as the colonizers and, by diverting their attention away from the class contradictions of British society, to feel patriotic pride in the achievements of the empire. In the 1930s and 1940s, "empire films" conveyed a message which depended on the cumulative effect of constant employment of certain viewpoints over successive generations and resulted in a colonialist set of meanings.[42] This vision of the relationship between colonized and colonizer had already been propagated throughout the nineteenth century, and "empire-cinema" itself had its roots in late Victorian and Edwardian travel films representing the colonized territories and embodying the same binary opposition between the colonial self, or the audience, and the colonized other, or subject of representation. Early British films of the colonies also conveyed an imperial message which was the result of a regular exposure to colonial stereotypes and certain viewpoints, and it thus helped to recreate India as a Western construct.

At the turn of the twentieth century, the cultural encounter between the British and India had already been long and intense. As Peter J. Marshall has argued, books about India and visual representations of the country and its people made by Western artists were widely available in Britain even by 1800, when the scale of contact between the two countries was greatly extended and an increasing number of people travelled to South Asia.[43]

41 Prem Chowdhry, *Colonial Indian and the Making of Empire Cinema: Image, Ideology and Identity* (Manchester and New York: Manchester University Press, 2000), 1.

42 Chowdhry, 1–5.

43 Peter J. Marshall, "Taming the exotic: the British and India in the seventeenth and eighteenth centuries", in *Exoticism in the Enlightenment*, 46.

Throughout the nineteenth century, visual representations of Indian cityscapes made by British observers had been popular in photography and paintings such as those by William Simpson.[44] Such visual representations of India contributed to the imperial project of scientific knowledge. As John Marriott has argued, metropolis and colony came to be understood through a belief in the principle of progress which took shape as India became subject to the knowledgeable gaze of the British state. British knowledge of India consisted of different forms of evidence such as historical inquiries, travelogues, mapping, censuses, libraries, museums and surveillance. These modalities brought India within the European imagination by gathering, classifying, categorizing and ordering information about the country.[45] They also gave rise, in the words of Ronald D. Inden, to "imperial knowledges" shaped by "universalizing discourses, the world-constituting cosmologies, ontology, and epistemologies produced in those complex polities at their upper reaches by those persons and institutions who claim to speak with authority."[46] As I have argued in previous chapters, museums and exhibitions marked Western dominance of the world by classifying, naming, displaying and mapping peoples and things from distant countries. India, as Gyan Prakash has argued, was an ideal focus for the pursuit of science and the collection of information. Indian exhibitions enhanced the importance of visuality in sustaining British rule in South Asia.[47] Displaced from their normal context, artefacts from India were exhibited in British cities and participated in the political and military domestication of India pursued through a surge of interest in the classification of knowledge.[48]

Cinematography promptly entered the system of overlapping images of India produced by the British for scientific and entertainment purposes.

44 See Mildred Archer, *Visions of India: The Sketchbooks of William Simpson 1859–62* (Oxford: Phaidon, 1986).

45 John Marriott, *The Other Empire: Metropolis, India and Progress in the Colonial Imagination* (Manchester and New York: Manchester University Press, 2003), 5–6.

46 Ronald D. Inden, *Imagining India* (Oxford: Basil Blackwell, 1990), 36.

47 Gyan Prakash, *Another Reason: Science and the Imagination of Modern India* (Princeton, NJ: Princeton University Press, 1999).

48 On Indian exhibitions see: Carol Breckenridge, "The Aesthetics and Politics of Colonial Collecting: India at World Fairs," *Comparative Studies in Society and History*, vol. 31, no. 2 (1989): 195–216.

Like nineteenth-century exhibitions in London, early films transformed unfamiliar and distant territories, making them comprehensible and tangible by displaying them for British audiences. Early travel films of India thus contributed to bringing Indian cityscapes within the British imagination, and they did so by focusing on everyday views of "native" individuals and activities in exotic and potentially threatening places. They put forward a domesticated and tamed image of Indian exoticism but broadly ignored political turmoil and the worsening of the Hindu-Muslim conflict caused by the partition of Bengal in 1905. In 1907, for example, the Walturdaw Company produced a series of films called *Missionary Life in India*. This series included scenes representing coconut plantations, Hindu and Muslim festivals – with no hint of the conflict between the two cults – and Buddhist sacred elephants, lepers and idol worship. "Native Street in India" is one of the most interesting scenes of this series because it represents a busy street scene in a town and offers a detailed view of everyday life. The camera is stationary and presents pedestrians and animal-drawn carts, low buildings and stalls on the left side of the street. No Europeans are visible, and the indigenous people in the scene seem to turn a slightly suspicious eye to the camera while pursuing their daily activities. The weather appears warm and sunny, and several men and women carry parasols. The scanty clothing of the "native" Indian men, most of whom are naked to the waist, marks a significant difference between British and Asian public space.[49] In Victorian and Edwardian Britain public nakedness was considered obscene and on public beaches men were supposed to wear modest swimsuits or use bathing machines.[50] At a time when prudery, as Richard Sennett argues, had its roots in the notion that "human meanings are immanent in all phenomena", the nakedness of indigenous people in urban settings violates the Victorian notion of modesty and it was made acceptable only by the cultural otherness of the subjects.[51] The theme of indigenous nakedness and dark skin

49 See W. Crook, "Nudity in India in Custom and Ritual", *The Journal of the Royal Anthropological Institute of Great Britain and Ireland*, Vol. 49 (1919): 251.

50 Bathing machines: roofed and walled wooden carts which would be rolled into the sea.

51 Richard Sennett, *The Fall of the Public Man* (London: Faber and Faber, 1986), 167.

colour had often featured in Victorian travel writing, and the exposure of the dark body on screen, as Ella Shohat suggests, further marked off boundaries between "self" and "other" by nourishing spectatorial desires.[52] Similar themes and meaning can be found in *Native Life in Ceylon*, a film which was produced by the Hepworth Manufacturing Company in 1909 and which has luckily survived. Although not officially connected with British India, Ceylon, presently Sri Lanka, was controlled by the British East India Company and had been declared a crown colony in 1802. The scenes filmed by Hepworth's cameraman recall those recorded by British filmmakers in India and, along with rural images of the inhabitants of a fishing village and Singhalese fakirs, include a market scene in Kandy and a street scene in Colombo. The former shows children, men and women dressed in traditional clothing in proximity to market stalls, where merchants have displayed their wares. The street scene in Colombo depicts ox-wagons with carriages, men and children sitting on steps and women passing by with parasols: like human subjects displayed in world's fairs and exploited to construct ethnographic surveys of the subcontinent, "native" people in Colombo are objectified in this cinematic representation of their living space. The presence of animals like chickens and goats contributes to making these representations picturesque to the eyes of Western audiences. Taking *Native Life in Ceylon* and "Native Street in India" as examples, the Western cameraman and audience can thus be identified as the receiver-observer of an environmental image wherein the identity of the place is an exotic urban landscape subdued to the observer's culture.

Imperial narratives of Indian urban spaces were displayed in a greater number of films made during the first two decades after the birth of the cinematograph. Few of these films have survived but their titles can help us to understand what kinds of travel films were popular in late Victorian and Edwardian Britain. In 1900, F.B. Stewart filmed *Our Eastern Empire*, a series of scenes for the Warwick Trading Company, including "Mid-day siesta in Calcutta", "Church Gate Street in Bombay", "Street Scene in Poona", and "Native street scene in Calcutta." Stewart, who was based

52 Shohat, 109.

in Poona in western India, was an experienced photographer who had already taken still photographs of British colonial and military life. His films seem to focus on "native" images of the Indian cities, thus contributing to collecting evidence for the British knowledge of India. In 1903, the Warwick Trading Company also produced a series of films representing the city of Hyderabad in the Indian state of Andhra Pradesh, which included "Panorama of street showing Hindu boy preaching" and "Street scene in Hyderabad", and which were filmed by Reverend J. Gregory Mantle, a Wesleyan minister and supporter of foreign missions. R.W. Paul produced *Scene in the Bazaar at Peshawar* in 1903 and Charles Urban made a series in 1908 called *India*, which included more "native" urban scenes such as "Mohammedan Festival at Calcutta", "Scenes in Benares, Delhi, Agra", "Madras", and "Through the native quarter, Bombay." In these films, which are not extant, just as in *Native Life in Ceylon* and "Native Street in India", the identity and meaning of the Indian city was strictly bound to the categories of the exotic and the picturesque which characterized earlier visual and written narratives of the colonies in Victorian Britain. The "natives" and the "native" landscape appeared as an exotic attraction for the use of Western audiences, and their identity was domesticated through a sensory experience which excluded political unrest, and the filth-ridden and poverty-stricken lives of the colonized peoples. In the following two sections, I look at these themes in films of Varanasi and Delhi.

Varanasi: A colonial cinematic narrative

In 1899 John Montagu Benett-Stanford, who is mostly remembered today for his films of the Boer War, made a series of films in India. The subjects of his films included the arrival of a train at Haora station in Calcutta, a crowded street scene in the same city and, the only surviving film, *Panorama of Calcutta*. Although always known by this title, the footage in fact appears to represent Benares, now Varanasi, one of the most sacred pilgrimage places for Hindus of all denominations. Filmed from a moving vessel on the river Ganges, this scene reveals a physical distance and cultural detachment between the cameraman and the subject of representation. It shows the Varanasi waterfront, pilgrims descending into the water, traditional Arab sailing vessels known as dhows and rafts. A large number of parasols and the partial nakedness of the dark-skinned "natives" also characterizes the scene, and the imposing structure of the embankment makes it a more monumental space than those depicted in "Native Street in India" and *Native Life in Ceylon*.

Described by Mark Twain as "older than history, older than tradition, older even than legend", a visit to Varanasi is revealed in this film as a partial and fragmentary sensory experience of urban life.[53] The film is a political description and narration related to earlier depictions of Indian cities made by British artists. Throughout the eighteenth and nineteenth centuries, texts, paintings and health maps, Swati Chattopadhyay suggests, entered the realm of imperial narration and, by settling the chaotic land and bringing it within colonial order, created a space of British authority that claimed to be the geographical space of India.[54] Indeed, Chattopadhyay's argument may also be applied to early actuality films: in order to explore and explain the Indian city, cinematic responses to it aimed at the creation of a safe and familiar experience of its exoticism. In films such as those taken by Benett-Stanford the picturesque performed an important role, creating a kind of visual mastery by suppressing the multi-sensory experience of the city, the disagreeable heat, sounds and smells, of the kind described by Josiah Conder who visited India in the 1830s:

> A constant creaking of cart wheels, which are never greased in India, a constant clamour of voices, and an almost constant thumping and jingling of drums, cymbals, & c...; and add to all this, a villainous smell of garlic, rancid coconut oil, and sour butter, and stagnant ditches.[55]

53 Mark Twain, *Following the Equator: a Journey Around the World* (Dover Publications, 1990; first published in 1897), 289.

54 Swati Chattopadhyay, *Representing Calcutta: Modernity, Nationalism, and the Colonial Uncanny* (London and New York: Routledge, 2006), 76–79 and 21–28.

55 Josiah Conder, *The Modern Traveller. A Description, Geographical, Historical and Topographical of the Various Countries of the Globe*, 30 vol. (Vol. III, London, Duncan, 1830), 49–50.

Early films such as *Panorama of Calcutta* helped constitute a stereotypical vision of the East but suppressed the most unfamiliar part of the sensory experience of the tropical city, in favour of a visual representation which focused on a domesticated image of colonial exoticism. *Panorama of Calcutta* thus contributes to the collection of evidence for the British knowledge of India, and is in this sense a perfect example of imperial narrative which aims at continuing part of the sensory experience of the city within the safe landscape and vision offered by the cinematograph.

Films thus served the same purpose as the domesticated sight of colonized individuals marching through the streets of London during events like Queen Victoria's Diamond Jubilee or displayed in venues such as the Crystal Palace or the South Kensington Museum, and discussed in Chapter 2. Narratives of national and colonial identity were often bound to capital cities and their sights and signs of power, and their social contradictions. As I demonstrate in the following section, the cinematic representation of Delhi differed significantly from that of other urban spaces in India.

Delhi the Durbar city on film

In early films of Delhi, before and after it became the capital of British India in 1911, the theme of the British rule emerged more clearly than in films of Varanasi and smaller towns. Two significant travelogues filmed in Delhi and recording daily life on the streets have survived: *Delhi*, which was produced by British Pathé in 1909 and presents a fascinating depiction of the city, and *Historic Mutiny Sites*, filmed in 1914 by a British cameraman.[56] *Delhi* represents the city centre during a Muslim festival, the 1908 Id-ul-Fitr.[57] The surviving print is incomplete, but it does include eleven shots of the festival. A panoramic view of the monumental old city with its temples establishes the location as ancient and historical, and is followed

56 Film credits are not extant.

57 The event celebrated the breaking the fast which follows Ramadan.

by a street scene with a crowd watching local street entertainers. A procession of people carrying shrines marches down the crowded and narrow street, while the following shot represents the monumental Jami Masjid, or Friday Mosque, which had been built by Shah Jahan in the middle of the seventeenth century. Several men are washing in the pool outside and then praying on the steps of the Mosque, while in the last scene, people are leaving through a monumental gateway. Cinematic images of ceremonies, as we have already seen, were often related to the most spectacular expressions of the empire, including parades and processions like the Diamond Jubilee. *Delhi* represents a tradition which did not have its origins in British imperialism, an indigenous Muslim crowd and event. The opposite is true for films of the Mutiny and the Durbars in Delhi.

Historic Mutiny Sites is a travelogue filmed around Delhi on the sites of the Indian Rebellion of 1857, or Sepoy Mutiny, the armed uprising which caused the end of the rule of India by the East India Company and the Mughal Empire, and their replacement by the British Raj – that is, direct rule by the British government.[58] The Sepoy Mutiny was an epoch-making event in the history of the British Empire. Its main consequence, as Bernard S. Cohn has argued, was the complete desacralization of the previous political and social order and of the person of the Mughal emperor.[59] Consequently it contributed, as John MacKenzie suggests, to the heightening of racial attitudes in imperial culture and to the atmosphere of hero-worship later characteristic of the age of high imperialism.[60] Newspapers amplified the general resentment caused by the Mutiny in Britain by describing the hideous actions perpetrated by the rebels in details. For example, the *Trewman's Exeter Flying Post* or *Plymouth and Cornish Advertiser* published accounts reported by the passengers of a ship, the *Colombo*, arriving directly from India to Southampton:

58 The rebellion started as a mutiny of sepoys, native soldiers allied to the Britain, on 10 May 1857 in the town of Meerut.

59 Bernard S. Cohn, "Representing Authority in Victorian India", in *The Invention of Tradition*, eds Eric Hobsbawm and Terrence Ranger (Cambridge: Cambridge University Press, 1983), 179.

60 MacKenzie, "Empire and Metropolitan Cultures", 280–282.

> The daughter of an English clergyman was driven through the streets of Delhi naked, then subjected to unspeakable outrages by an infuriated soldiery, and afterwards cut to pieces with swords. An English lady in the same city was suspended by the feet naked and hacked to pieces.[61]

Indian places and events were prominent among the subjects of mid-nineteenth century shows such as the moving panoramas. The Indian Mutiny was staged in London in popular plays and, immediately after its suppression, was the subject of a show at Burford's Panoramas in Leicester Square. The show included the views *Fall of Delhi* and *The City of Lucknow*.[62] The shocking effect of the Mutiny in the imperial system was also reflected in visual cultures of the empire, and fitted well the kind of material recorded in early films.

Even during the Mutiny itself, the events in India were the subject of visual representations in cartoons and drawings published in British newspapers and displaying images of Indian brutality which reflected the sense of British national outrage.[63] Indeed, the mutiny remained a popular subject of representation and early cinematography celebrated its suppression on its fifty-eighth anniversary in 1914: the film *Historic Mutiny Sites*, as the title suggests, represents places in which the decisive events unfolded, as well as everyday activities of the "native" population in the streets of Delhi. The film is opened by a medium shot of a perimeter wall of the city with horse- and ox- traffic passing through a double gateway and this is followed by a close-up of one of the archways. Dignitaries arrive at the train station and gather in a garden, Indian tailors work sitting cross-legged on the city pavements, and crowds walk up and down a stone staircase in the central district of the city. These images are followed by scenes exemplifying British patriotism and the suppression of the Mutiny: a statue of General John Nicholson (the so-called Hero of Delhi), the Mutiny Memorial built on the site of

61 *Trewman's Exeter Flying Post or Plymouth and Cornish Advertiser* (Exeter, England), 13 August 1857.

62 See Richard D. Altick, *The Shows of London* (London and Cambridge, MA: Harvard University Press, 1978), 177.

63 Denis Judd, *Empire: the British Imperial Experience from 1765 to the Present* (London: HarperCollins, 1996), 66–70.

the British Camp in 1857 on the northern ridge, St James's Church and the Delhi Fort. Nicholson played a legendary role in the Indian Mutiny as he planned and led the storming of Delhi: the statue marked the place where he fell. The Fort was used by the sepoys as the Indian headquarters, and was eventually occupied by the British troops of General Nicholson. Finally, we see a plaque on the so-called Magazine Gateway, which commemorates the six soldiers of the empire who destroyed a deposit of gunpowder and themselves in order to evade capture. The Mutiny had created profound fear and distrust of Indians in Britain. In 1914, almost sixty years later, the suppression of the Indian rebellion had became a glorious imperial memory and *Historic Mutiny Sites*, as we have seen, celebrates this event by filming the sites of the battle, which were by then inscribed into the monumental space of the city of Delhi. This montage thus contained a direct political purpose, promoting the British colonial adventure in India by presenting a view of the city as subdued to the empire.

An even more imperialist significance emerges from the films of the Delhi Durbars marking the coronation of Edward VII in 1903 and George V in 1911. These events regenerated an Indian tradition and, as David Cannadine argues, consolidated and completed the British-Indian hierarchy with inflated and extravagant ceremonies.[64] On Saturday 17 January 1903, *The Illustrated London News* devoted its cover to a watercolour representing the building of a triumphal arch at the Mori Gate. A drawing representing the viceroy's state procession during the Durbar was also published in that issue, and two photographs of a mosque in Delhi and a gateway in the Parana Khila, or the Old City, were accompanied by a title defining Delhi as "The Durbar City".[65] Narayani Gupta has effectively explained how great the influence of the Durbars was on the life of the city:

64 David Cannadine, *Ornamentalism: How the British Saw Their Empire* (London: Penguin Books, 2001), 41–57.

65 *The Illustrated London News*, No. 3326, Vol. CXXII (Saturday, January 17, 1903): 78, 83, 102, and 104.

> During the Durbars of 1903 and 1911, those extravaganzas when the Imperial Government cast its gigantic shadow over the local, a city momentarily came into existence adjacent to the living organism of Delhi. Electricity and tramways and railways lines transformed acres of fields into settlement described as being the size of greater London in 1903 and one many times larger in 1911.[66]

Delhi flourished in the second half of the nineteenth century. In the 1890s it became the commercial capital of the Punjab and it was to become the political capital of British India in 1911. The city saw an increase in population from 154,000 in 1868 to 206,000 in 1901, and a transformation of its appearance as a result of the development of a western suburban area (from Mithai Ka Pul to Jhanderlop), and the shifting of the Cantonment to the Ridge. This transformation, as Narayani Gupta has explained, led to the extension of municipal facilities: electricity came to Delhi in 1902 as a side-effect of the Durbar, and thanks to municipal funding and private philanthropy, a tramline network was introduced for the purpose of linking the city to and travelling within the suburbs.[67]

The Coronation Durbars of 1903 and 1911, according to Alan Trevithick, were political rituals, or ideological and aesthetic social constructions aimed at legitimizing and popularizing British rule in India by employing symbolic action. The intent of the Durbars, Trevithick has argued, was to reach out to the large illiterate masses by influencing the targeted audience, or the "native" minds, of India.[68] Just as the Indian masses were thus targeted by the organizers of the Durbar, British audiences were the target of the Durbar films, whose purpose was to promote a sense of common participation in the colonial political system. Like films of the Diamond Jubilee procession, the coronation of Edward VII, and other imperial parades and colonial exhibitions in Britain, the Durbar films contributed to making British citizens familiar with a domesticated image of colonial people. They also contributed to the consolidation of an imperial system based on

66 Narayani Gupta, *Delhi Between Two Empires 1803–1931: Society, Government and Urban Growth* (Oxford and Delhi: Oxford University Press, 1981), 175.

67 See Gupta, 157–185.

68 Alan Trevithick, "Some Structural and Sequential Aspects of the British Imperial Assemblages at Delhi, 1911", *Modern Studies*, vol. 24, no. 3 (1990), 561–562, 573.

a distinction between concepts such as colonizers, home and domination on one hand, and colonized, exoticism and submission on the other hand.

Trevithick has explained how the 1903 Durbar was prepared by Lord George Nathaniel Curzon, Viceroy and Governor-General of India, with the purpose of promoting a sense of common participation in the imperial political system by bringing the new sovereigns into communication with his people in a solemn public ceremony. In particular, the attempt by Lord Curzon to embed Indian traditional features into the texture of the Durbar was aimed at emphasizing the continuity between Mughal and British rule, and the legitimacy of the latter, by breaking through the sensitive level of symbolic receptivity in the "native" mind.[69] The programme of events began on 29 December 1902 with a procession through the streets of Delhi, and included ceremonials, sports, music and competitions. R.W. Paul, the Warwick Trading Company and Gaumont all sent their cameramen to film the Durbar and, in particular, the procession on 29 December, and several fragments of the films produced by those companies have survived. Warwick's *The Delhi Durbar* is the only surviving film to record the Durbar celebrations on the streets of Delhi: it represents the crowded streets of the old city on the day of the procession. The celebration invested everyday life in the city with a sense of imperial power and submission to the British crown through the re-enactment and renewal of local, pre-colonial, rituals and customs. Its cinematic representation amplified these meanings and brought them before the British masses more effectively than other media. As Stephen Bottomore has argued, the cinematic image of the 1903 Durbar exemplified the European perception of India as exotic visual spectacle and soft propaganda for British rule in India: cinema, as a medium which itself thrived on spectacle and display, thus spread around the world an image of Indian costume orchestrated by the British authorities.[70] The Durbar celebrations were staged for the Indian masses which experienced

69 Trevithick, 567–568.

70 Sthephen Bottomore, "An Amazing Quarter Mile of Moving Gols, Gems and Genealogy: Filming India's 1902/03 Delhi Durbar", *Historical Journal of Film, Radio and Television*, vol. 15, no. 4 (1995), 511.

the event on the crowded streets of Delhi but, like the film relating to the 1857 Mutiny, they were distributed and exhibited in Britain's music halls: the Gaumont film was shown at the London Hippodrome, the Empire and the Egyptian Hall; the Warwick films were shown at the Palace and Alhambra theatres in London; Paul's films were also shown in Leicester from 25 January 1903.[71]

Similar considerations can be applied to the 1911 imperial Durbar and its films.[72] This Durbar was constructed around the actual presence of George V and the movement of the Indian capital from Calcutta to Delhi itself, which was presented as the traditional seat of the Indian Empire. As Robert E. Frykenberg has argued, the decision to abandon Calcutta followed an abortive attempt to divide the province of Bengal into two zones and dealt a blow to Bengal's embryonic national movement.[73] In 1911, King George V and Mary of Teck, the Queen Consort and Empress of India, went to Delhi in person, but the Indian traditional elements were not as prominent as in 1903. The 1911 Durbar was one of the biggest events to be filmed since the Diamond Jubilee in 1897, and it was filmed by over twenty companies, including Gaumont, Urban, Pathé, Warwick Trading Company, and Baker Motion Photography. The Warwick Trading Company had previously filmed and distributed a scenic film of Delhi, which was described in the company catalogue as a portrait of "the beauties of the ancient city in which the Great Durbar will take place."[74] Prints of the films made by the Warwick Trading Company, Gaumont, and Barker Motion Picture have survived and they represent scenes in the huge arena

71 Bottomore, 510.

72 The first Durbar celebrated the proclamation of Queen Victoria as the first Empress of India in 1877.

73 Robert E. Frykenberg, "The Coronation Durbar of 1911: Some Implications", in *Delhi Through the Ages: Essays in Urban History, Culture and Society*, ed. Robert E. Frykenberg (Oxford and Delhi: Oxford University Press, 1986), 369–390. See also A. Volwahsen, *Imperial Delhi: The British Capital of the Indian Empire* (Munich, Berlin, London and New York: Prestel, 2002), 11–16.

74 See Stephen Bottomore, "Have You Seen the Gaekwar Bob?: Filming the 1911 Delhi Durbar", *Historical Journal of Film, Radio and Television*, vol. 17, no. 3 (1997): 326.

at the Durbar camp during the proclamation of King George V as Emperor of India on 12 December. Baker Motion Pictures also filmed the Calcutta Pageant held on the afternoon of the 5 January 1912 as part of the celebrations for the coronation of George V. On the following day, the Durbar films were screened for the royal couple at Government House in Calcutta: *The Illustrated London News* recorded this event on the cover of its issue of Saturday 3 February, which featured a drawing of the royal couple watching the Durbar films (just as Victoria had entertained herself at Windsor Castle with films of her own Jubilee in 1897).[75] As in 1903, films of the George V Durbar were widely distributed and exhibited, and the Gaumont film was first shown in London at the Electric Palace, Marble Arch, on Saturday 30 December 1911. During the weeks and months that followed, the Durbar films were shown all over Britain and the empire.[76]

Both the 1903 and 1911 films of the Delhi Durbars displayed imperial propaganda built around three main themes: the exoticism of India, its restored and renewed past, and the British presence in the country. The word "Durbar" (originally Persian for the Shah's noble court) was used in India for the court of "native" rulers of Mughal who were said to receive visitors in audience and conduct business *in durbar*: the British monarchs had long since replaced the Mughal emperors at the summit of the social order and the Durbar celebrations confirmed their rule. By adopting the word "durbar" to define ceremonials visually influenced by Indian traditions, the organizers sought to address and engage the Indian masses. Like *Delhi Mutiny Sites*, films of the Durbars thus displayed these three themes exemplified in the large but confined space of the Durbar ceremonies. Exoticism and British rule took the form of images of traditional costumes and ceremonial actions, such as the use of numerous elephants, and on the other hand resulted in images consolidating the British Raj by recording the presence of British dignitaries and even the King and the Queen themselves. The "native" images of Delhi during Id-ul-Fitr filmed in British

75 *The Illustrated London News*, No. 3798, Vol. CXL (Saturday, February 3, 1912).

76 See Bottomore, "Have You Seen the Gaekwar Bob?: Filming the 1911 Delhi Durbar", 325–329.

Pathé's 1908 film *Delhi* were thus replaced in the Mutiny and Durbars films by a British imperial image of a city fitting its role of capital of the British Raj and exemplified by images and memories of imperialist victory and its ceremonial consolidation. Indian masses were the audience of the Durbars and, along with all other citizens of the empire, the audience of films where Delhi became an urban symbol of British imperialism in India.

This imperialist image of Delhi, as we have seen, also differs from other representations of urban space such as the films of Colombo and Varanasi. These early films, just as photography and colonial exhibitions, transported the spectators to an India existing in part as the product of unchecked British imagination. Late Victorians and Edwardians stepped into this exotic world through different media, but cinematic exposure to the East would soon dominate visual culture and consolidate the East as a product of European imagination. Only one other territory could rival India in terms of exoticism and romanticism, and even surpass it in terms of unknown danger and mystery, in the imagination of the British. Only partially or indirectly subdued by European empires, China and its trading port cities played a significant role in British visual culture, including in early travel films.

CHAPTER 7

The Perilous Exoticism of the East

Where we encounter allegedly righteous and harmonious fists, an unlikely alliance and a Conservative MP visiting the Far East with a movie camera.

As much as the Indian exotic, British-made images of China invoked the idea of cultural otherness. In 1911 Charles Urban produced a series of films representing China. One of them, *In Quaint Peking*, has survived and offers an interesting representation of "native" everyday life in Peking. The first scene shows the crowded South Gate, its chaotic traffic of carts and rickshaws. Signs written in Chinese logograms are visible, as are clothing stores and labourers delivering goods. The "natives" look diffident and stare at the camera with suspicion. The following shots represent a local hairdresser, a shoemaker and a Chinese lady smoking her pipe. This is followed by a panorama of the banks of the Guanting Reservoir representing manual labourers during their lunch hour, with a large number of junks and sailing boats in the background. *In Quaint Peking*, like other actuality films of cities produced between 1899 and 1914, focuses on "native" views and everyday activities and reproduces the East as an alien yet picturesque place for the amusement and education of its audience. In this chapter, I look at early films of cities in mainland China as well as of Crown colonies such as Hong Kong and Singapore, examining the exoticization of those locales and their people, as well as defining historical events such as the Boxer Rebellion of 1900.

China played a significant role in the British imagination of the Orient and her cities were often represented by British artists and entertainers. The South Kensington Museum in London – as I mentioned in Chapter 2, itself the subject of cinematic representations in 1901 – exhibited an impressive collection of Chinese art. Popular forms of entertainment were also concerned with China and its people. For example, Burford's panoramic canvases in Leicester Square, where the Sepoy Mutiny had also been depicted, often represented events related to the British campaign to open China to western trade.[1] Canton was portrayed in 1838 and in 1859–1860 in a series of views advertised as taken "from the Magnific Hill showing the City and beautiful surrounding scenery."[2] Macao was the subject of a show in 1840, and Hong Kong and Nanking in 1844–1845. These shows were advertised in the newspapers with an emphasis on their exotic and their patriotic characteristics. In November 1840, *The Era* published this advertisement for the panorama of Macao:

> CHINA – LEICESTER SQUARE – BURFORD'S splendid PANORAMA of MACAO, with the Bay of the Typa and the numerous picturesque islands by which it is surrounded. In the Bay are represented the Volage and Hyacinth, and others of her Majesty's ships, the Tender of Captain Elliot, Chinese junks, and numerous small native craft of singular construction.[3]

The panorama of Nankin, as *The Satirist* commented, drew "many visitors out of the vast multitude who must have heard much, but seen little, of the Celestial Empire."[4] China, as Burford's shows demonstrate, occupied a central place in nineteenth-century popular visual culture. On a more political level, the British presence in the East Asia, as Jürgen Osterhammel argues, was more than a metaphor for commercial supremacy in the China Seas, and was particularly strong in comparatively modern urban environments created by the European invasion itself, such as Hong Kong, Shanghai and Singapore, with their enclosed colonial societies. Osterhammel has explained how during the nineteenth century a system of international treaties was established that turned large parts of China into an uncolonized extension of the British Empire:

1 Altick, 177.
2 *The Morning Chronicle* (London, England), Friday, April 8, 1859.
3 *The Era* (London, England), Sunday, November 22, 1840.
4 *The Satirist; or, the Censor of the Times* (London, England), 27 April 1845.

> The treaties guaranteed rights of access to and of residence in a number of major Chinese cities (transformed into treaty ports), personal security of foreign citizens from the alleged "barbarity" of Chinese justice, a uniformly low tariff, and a privileged treatment of foreign goods in transit through the customs-ridden Chinese interior. They also opened up China's rivers and coastal waters to the unchecked activities of foreign shipping companies.[5]

The British campaign to open China to western trade did not diminish the charm and mystery surrounding China and its cities, which remained a source of excitement for the British imagination and the subject of a number of prints, drawings, photographs and films. In the early 1870s, Scottish photographer and geographer John Thompson, Fellow of the Royal Geographical Society, travelled to Cambodia, Malaya and China, taking photographs to document the life of people, landscapes, cities and artifacts. Thompson, who later cemented his reputation taking photographs of the streets of London, was one of the first Victorian photographers to travel to the Orient and his nine books of photographic exploration in eastern and south-western locations in Asia were among the most remarkable photographic documentation of their time. Thompson's work includes *Illustrations of China and Its People*, which was the pinnacle of a long-standing fascination and esteem surrounding the image of China in Britain during the first part of the nineteenth century when the Chinese were praised for their exotic works of art. This esteem was also accompanied by a contrasting feeling of fear. From the mid nineteenth century, as Catherine Pagani has argued, two different attitudes towards China coexisted in Britain. The British victory in the Opium War (1839–1842) diminished the esteem previously felt for China and led to the stereotyping of the Chinese in Britain as a lawless race. China was thus perceived as the land of the uncivilized and the country of the exotic.[6] The Chinese were admired and increasingly feared and despised. This ambivalence towards China

5 Jürgen Osterhammel, "Britain and China", in *The Oxford History of the British Empire: The Nineteenth Century*, vol. III, ed. Andrew Porter (Oxford and New York: Oxford University Press, 1999), 146–149.

6 Catherine Pagani, "Chinese material culture and British perception of China in the mid-nineteenth century", in T. Barringer, and T. Flynn, *Colonialism and the Object:*

produced a number of orientalist clichés and racial stereotypes which were constantly projected upon China in writing and visual culture.

At the turn of the twentieth century, when British filmmakers followed the path of John Thompson and recorded moving pictures of China and its people, China still held a special place in the Western view of the Orient. At that time, the image of China as an uncivilized land was strengthened by the Boxer Rebellion, an anti-foreign, anti-imperialist Chinese uprising from November 1899 to September 1901, which began as a peasant-based movement in the northern part of the country. The Boxers, or Society of Righteous and Harmonious Fists, attacked foreigners and the Christian Chinese alike, killing thousands in Beijing and in Shandong and Shanxi Provinces. Britain took part in the Eight-Nation Alliance, a campaign which put down the Boxer Rebellion and forced the imperial government to sign the Boxer Protocol of 1901.[7] The Boxers' abhorrent acts of looting, murder, and rape were often recorded by the British press, thus strengthening the perception of the Chinese as a lawless race. On the other hand, as Mike Davis has argued, at least 2 million Chinese had starved to death in the two years prior to the rebellion and many Chinese believed that that the disaster was man-made, caused by Westerners' construction of missions, churches, and cathedrals which disrupted the geomantic balance of nature, causing floods and drought. These events contributed to the transformation of the Boxer Uprising into a broad-based popular movement and, according to Davis, explained its brutality to a great extent.[8] Davis has also shifted the focus from the undeniable brutality of the Boxers to the equally undeniable, but often overlooked, violent actions committed by the Eight Nations armies. Early films of China did not engage with the representation of famine or its consequences, or with the brutal actions of the Western armies. Rather, they created meaning by their absences,

---

Empire, Material Culture and the Museum, London and New York: Routledge, 1998, pp. 28–40.

7 See Diana Preston, *The Boxer Rebellion: The Dramatic Story of China's War on Foreigners that Shook the World in the Summer of 1900* (New York: Penguin, 2001).

8 Mike Davis, *Late Victorian Holocausts: El Niño Famines and the Making of the Third World* (London and New York: Verso, 2001), 178–188.

enforcing an image of the uprising which defines a confrontation between what Western audiences would have perceived as the civilized West and the barbarian East, and thus consolidating the perception of a "yellow peril."

This view pervaded all aspects of popular culture. In the autumn of 1900, the magazine *The Halfpenny Marvel* published *The Curse of China*, a work of fiction tackling the Boxer Rebellion in which "missionaries, British and others would be sacrificed to the mad, religious hate of the yellow man."[9] From the early 1910s, fear of the Chinese would be the theme of the popular Fu Manchu novels by Sax Rohmer and it also influenced early cinematography. The Boxer Rebellion also provided the subject of one of the most remarkable fictional films of the early twentieth century, James Williamson's *Attack on a China Mission* (1900), which focused on the Boxers' attempt to expel Christian missionaries, representing a mission under attack and its rescue by Royal Marines. Williamson's film was one of the most successful films of its time, and has often been praised for its innovative editing and well-developed narrative.[10]

Attack on a China Mission film was actually filmed by Williamson in Hove, Brighton, rather than in China itself. Several fake actuality films representing the Boxer Rebellion were produced by British filmmakers (just as, fake re-enactments of the Boer War were also popular in the early days of cinematography, as we will see in the next chapter). For example, Walter Gibbons produced a series of staged films representing the atrocities perpetrated by the Boxers: *The Burning of a Missionary*, which represents Chinese soldiers hanging a missionary upside down over a fire, and *Boxers Decapitating a Prisoner*. These films served as an even more partial representation of the events focusing on the ruthless actions of the Boxers, but omitting episodes of misconduct by the British troops. Gibbons also made a few films on location, recording actual events in China: *Outside the Walls of Peking*, which shows Allies capturing Boxers, and *Allied troops taking Chinese Prisoners over Tientsin Bridge*.

9 *The Halfpenny Marvel* (London, England), 6 October 1900.
10 See Burch, 107.

A small number of actuality films recording events in Chinese urban spaces were filmed by other British filmmakers. In 1900 the Conservative MP Ernest Hatch travelled to East Asia and filmed with an unknown cameraman about twenty actuality films which he later distributed in London with the Harrison Company. Hatch's *Volunteer Turnout at Shanghai*, now lost, recorded the actions of volunteers protecting foreign settlements from the Boxers, and was in this sense a factual counterpart of Williamson's *Attack on a China Mission*. Hatch also filmed everyday activities in *Street Scene in Peking* and *Panorama of Hong Kong Harbour*, and he wrote a book about his journey, *Far Eastern Impressions*. Although, unfortunately, he does not discuss the filming, the book does contain a number of photographs taken by Hatch, including several pictures of street scenes in Peking depicting funeral and wedding processions and street cleaning.[11] Hatch's written description of Peking reflects the ambivalence of attitudes towards China in Britain and reveals the exotic expectations of the traveller:

> Despite its historic interest, Peking is one of the most disappointing cities that I have ever visited. It seems to typify all the vices and weakness of modern China. It is squalid to a degree inconceivable to one whose acquaintance is only with populous places of the West; the public buildings are conspicuously mean-looking and there is an almost entire lack of picturesqueness which lends to most Eastern cities a charm to the eye of the traveller from the West. The filth everywhere is astounding. Sanitation is quite unknown and most elementary conservancy is neglected.[12]

Other British filmmakers turned their attentions towards Eastern cities and the exoticism which lent them a charm to the eye of the Western traveller. In 1904, Harold Mease Lomas undertook a filming trip to British North Borneo and filmed for Charles Urban *The Urban-Bioscope Expedition into Borneo*, a series of forty-four scenes including urban scenes such as "Native and Chinese Quarters, Sandakan", and "A ramble through Singapore." Arguably the most remarkable example of travel cinematography in the

11 Ernest Hatch, *Far Eastern Impressions* (London: Huchinson, 1905). These are the titles of the photographs: *A native funeral*, *A view from the wall*, *A scene in Peking*, *A busy scene outside the gates*, *A wedding procession*, *Watering the road*, and *A street scene*.

12 Hatch, 150.

Orient at the beginning of the century, however, was to be found in the films made by Joe Rosenthal for the Warwick Trading Company in 1900. Rosenthal filmed everyday activities in Singapore, Tientsin, Shanghai, Peking, and Canton (now Guangzhou).[13] *Panorama of Singapore Sea Front* and *Circular Panorama of Singapore and Landing Stage* are the only two films made by Rosenthal in the Crown colony of Singapore during his journey.[14] Tientsin (presently Tianjin) was the subject of scenes representing its "native" quarters and its lively streets, and was also depicted in a panorama filmed from the Hai He River. Two street scenes focused on the occupation of Tientsin by allies following the Boxer Rebellion: *Street in Tientsin after Occupation by Allied Troops* and *Streets and Ruins of Tientsin*. Peking was also filmed with a focus on the action of the troops of the Eight-Nation Alliance in *Streets of Tientsin after the Town's Capture*, *American Transport Entering Peking*, and *Entry into Peking of Count Von Waldersee*, which recorded the arrival of the supreme commander of the joint forces in China Alfred Graf Von Waldersee (who actually arrived in Peking too late to direct his troops in fighting). Other scenes of Peking included a panorama filmed from the Tartar Gate and a camel transport entering the city. Rosenthal's films of Shanghai and Hong Kong focused almost exclusively on their "native" populations and spaces. The latter had

13 Titles: "Busy Street in Shanghai", "Sikhs' Camp in Shanghai", "Chinese junks in Hong Kong Harbour", "Circular Panorama Hong Kong Harbour", "Chinatown Bazaar – Hong Kong", "Ride on the Peak Tramway", "Lively Stret Scene in Shanghai", "Curious Natives on Shanghai Streets", "Circular Panorama of Native Quarters", "Foreign Warships off the Bund", "Panorama of the Bund", "Chinese Market and Canal", "Street in Tientsin after Occupation by Allied Troops", "Streets and Ruins of Tientsin", "Native Quarters of Tientsin", "Panorama of Tientsin from the River", "Street Life in the Tartar City", "Panorama of Peking from the Tartar Gate", "Camel Transport Entering Peking", "American Transport Entering Peking", "Entry into Peking of Count Von Waldersee", "Jinricksha Stand – Shanghai", "Sea Front Drive of Shanghai", "Nankin Road, Life in a Canton Street", "Queen's Road – Hong Kong", "Shanghai from a Sailing Junk", "Shanghai Shops and Opium Dens", "Review of the Fourth Goorkhas at Hong Kong" and "Streets of Tientsin after the Town's Capture."

14 Urban's scene "A ramble through Singapore" and *Singapore*, a British Pathé production from 1910, are arguably the only other films of Singapore from this era.

been a British colony since 1842, and was the subject of a series of scenes representing its harbour with Chinese junks and foreign warships, Queen's Road, a Chinese Bazaar, and the Bowrington Canal (Wong Nai Chung).

The British settlement in Shanghai, as Osterhammel has argued, was the most important asset of the British Empire in East Asia and a crucible of economic and social modernization.[15] It was also the subject of a number of scenes filmed by Joe Rosenthal and representing its shops and opium dens, its crowded streets, its lively "native" quarters and its Sikh Camp, coolie boys diving for coins, and Japanese style rickshaws known as Jinricksha. The only film of Shanghai known to have survived is *Nanking Road, Shanghai*, which offers a representative case. This film depicts Shanghai as an urban space where different cultures are represented by foreign presences on its streets. Nanking Street was, and still is, a busy commercial thoroughfare. Later renamed Nanjing Road, it epitomizes the creation of a modern commercial culture which was central to the transformation and modernization of China's economy.[16] In Rosenthal's film it appears crowded with people pursuing their everyday activities, shopping or working. The "native" look of this street is given by the presence of traditional rickshaws, sedan chairs and a four-wheeled vehicle, and a number of posters and signs written in Chinese logograms. A European lady rides a bicycle while two German officers, part of the Eight Nations Alliance troops, walk down the road. Indian subjects of the British Empire were often recruited for police service; in this film we can see detachment of Sikhs walking down the street.

The Boxer Rebellion influenced the cinematic image of China, both in films representing troops and rebels, and in street scenes like *Nanking Road, Shanghai*. Chinese artefacts brought back to Britain after the campaign, and stories of unspeakable cruelties perpetrated by the rebels, fuelled the ambivalence with which China was perceived in Britain as the land of both the uncivilized and the exotic. The Boxer Rebellion defined British perception of China just as the Boer War defined the British perception

15 Osterhammel, 148.

16 See Sherman Cochran (ed.), *Inventing Nanjing Road: Commercial Culture in Shanghai, 1900–1945* (Ithaca, NY: Cornell University Press, 2000).

of Africa. In the following chapter, I shall focus on the ways in which the Boer War influenced the cinematic image of urban space in the territories which would later become known as the Union of South Africa, and I shall then turn my attention to films of British dominions in North Africa and the Middle East.

CHAPTER 8

From the Boer War to the Great War

In which early propaganda films record a British victory in Africa while anticipating larger, more devastating wars and, with them, the end of the Empire.

Images of Africa fueled the geographical imagination and knowledge of the Victorians and the Edwardians, in ways that present both elements of continuity and discontinuity when compared to images of Asia. People and artefacts from the African continent were often displayed in ethnographic exhibitions and museums. As Annie Coombes argues, images of Africa at large-scale national and local exhibitions had a tremendous impact on the popular imagination and served to aggrandize the imperial message.[1] Films brought this message to a larger audience by filming the African exhibitions held in Britain.

In 1899, the show "Savage South Africa" opened at the Greater Britain Exhibition in Earl's Court, London, and included two hundred "natives" from Africa, twenty Boers, and the re-enactment of episodes from the Matabele Rebellion in Zimbabwe.[2] Several films were shot at Earl's Court during the exhibition. G.A. Smith and John Benett-Stanford filmed *Kaffirs and Zulus on the Warpath* and *Savage South Africa at Earl's Court*. The Warwick Trading Company and Walturdaw produced *Opening of the South*

1 Annie Coombes, *Reinventing Africa: Museums, Material Culture and Popular Imagination in Late Victorian and Edwardian England* (New Haven and London: Yale University Press, 1984).

2 See Mai Palmberg, *Encounter Images in the Meetings Between Africa and Europe* (Uppsala, Sweden: Nordic Africa Institute, 2001), 71–72.

African Exhibition and *South African Products Exhibition* respectively, two actuality films representing the South African Exhibition and thus emphasizing the connection, and at some degree the submission, between the Union of South Africa and Britain.[3] At Earl's Court, the Warwick Trading company also produced *Savage South Africa: Savage Attack and Repulse*, a scene from Frank E. Fillis's "Savage South Africa", representing British infantry and cavalry during an attack by African "natives", and performed during the exhibition at the Empress Theatre in Earl's Court. As much as the exhibitions they portrayed, these films documented imperialism as the dominant ideology and exemplified the ethnocentrism of early cinematography. Much the same can be said of travel films shot in Africa during the same era which offered the opportunity to their audiences to take part as observers of newly reimagined communities in the imperial system.

The image of a self-contained imperial world emerged from several films of African cities and landscapes. Between the Greater Britain Exhibition and the South African Exhibition, Britain fought a war for the domination of South Africa. This event marked the cinematic representation of cities such as Cape Town, which were normally portrayed as war zones and, after victory, as colonial settlements. By contrast, exoticism and native spaces were still a characteristic of the cinematic image of cities in North Africa. Fascination with the African continent was particularly strong in late Victorian Britain and, as we will see in this section, was displayed in a great variety of early films whose settings were geographically dispersed, and which recorded nature, the built environment and the people of Africa. Filmmaking, like longer-established activities pursued in the colonies, including photography and the hunting and display of wild animals as skin trophies, all involved representing life in Africa and objectified their colonized subjects. Images of Africa served as imperial propaganda through direct support for colonizing forces and through the

3 Both films are now considered to be lost. The latter was a series including "Acrobatic Piccaninnies", "Piccaninnies and Lion Cubs", "Taming a Lion", "Parades of the Apes", "Boers Hunting Zebras and Wild Beasts", "The Somersault Ape", "Captain Taylor and His Trained Cinnamon Bear", and "Savages attacking the Square".

inclusion and omission of certain themes and subjects.[4] Early filmmaking in Africa is thus an historical reflection seen through the eyes of the society which produced and consumed the films as agents and products of an imperialist agenda.

The Boer War and South African cityscapes on film

Just as the Boxer Rebellion had influenced the filming of China, at the turn of the twentieth century both fictional and factual cinematic representations of South Africa were dominated by the ongoing Boer War.[5] Fought from 11 October 1899 to 31 May 1902, between the forces of the British Empire and the two independent Boer republics of the Orange Free State and the South African Republic, this was the most expensive British military expedition since the Napoleonic wars. On the one hand, it was caused by a refusal by the Boers to offer political rights to the mainly British *Uitlander* immigrant workforce. On the other hand, it was caused by Britain's imperial ambitions in South Africa, rich in gold and diamonds, and its rivalry in the region with Germany. An initial Boer invasion of Natal, in which British forces were beaten back, was repelled by British troops led by Sir Redvers Buller, Lord Frederick Sleigh Roberts and Lord Horatio Herbert Kitchener, and ensued a subsequent guerilla war. Britain eventually won, but at the high cost of over 20,000 lives.[6]

4 See, for example, Paul S. Landau, "Empires of the Visual: Photography and Colonial Administration in Africa", in Images and Empire: Visuality *in Colonial and Postcolonial Africa*, ed. Paul S. Landau (Berkeley: University of California Press, 2002), 140–171.

5 See Thelma Gutsche, *The History and Social Significance of Motion Pictures in South Africa, 1895–1940* (Cape Town: Howard Timms, 1972).

6 See, for example, Thomas Pakenham, *The Boer War* (London: Macdonald, 1979); Denis Judd and Keith Surridge, *The Boer War* (London: John Murray, 2002); and Donal Lowry (eds), *The South African War Reappraised* (Manchester: Manchester University Press, 1999).

As Simon Popple suggests, cinema was crucial to the construction of discourses surrounding the Boer War.[7] As we have seen in the previous chapters, the war had a great impact on public opinion in Britain, where it consolidated patriotic belief in the empire. This impact was reflected in the number of films shot in London, Liverpool, Manchester and Southampton which celebrated the troops leaving Britain or returning from South Africa, and in images of hero-worship of the commander Lord Roberts. Richard Price has suggested that the public excitement for the war in Britain was essentially a middle-class phenomenon and, despite an increase in the working-class response to recruitment during the war, Britain's workers generally focused on domestic concerns.[8] Notwithstanding this, early films of the Boer War brought this event closer to the working-class audiences of the music halls, where images of the military expedition became increasingly popular. These films were distributed and exhibited in Britain and attracted the audience with a promise of realism. On the other hand, as we shall see in the last section of this chapter, in Ireland the brutality of the Boer War invigorated the Irish nationalist movement.

At least eight British cameramen are known to have filmed in South Africa during the Boer War: William Kennedy-Laurie Dickson for the British Mutoscope and Biograph Company, Walter Calverley Beevor and Sydney Melsom for Robert W. Paul, John Benett-Stanford (who had already filmed the Sudan campaign in *Alarming Queen's Company of Grenadier Guards at Omdurman*, 1898), Edgar M. Hyman and Joe Rosenthal, who was replaced by Sydney Goldman in 1900, for the Warwick Trading Company, and C. Rider Noble for Walter Gibbons.[9] Other filmmakers, such as G.A. Smith, resorted to "fake" recreations of battlefield scenes filmed

7 Simon Popple, "Startling, realistic, pathetic: The Mitchell and Kenyon 'Boer War' Films", in *The Lost World of Mitchell and Kenyon: Edwardian Britain on Film*, 150.

8 See Richard Price, *An Imperial War and the British Working Class: Working-Class Attitudes and Reactions to the Boer War, 1899–1902* (London: Routledge and Kegan Paul, 1972).

9 See John Barnes, *The Beginning of Cinema in England, 1894–1901: Filming the Boer War*, London: Bishopsgate Press, 1992.

in Britain.[10] Mitchell and Kenyon also produced a series of reenactments of battles, individual heroic actions by British servicemen, and atrocities committed by the enemies of the empire. These films were to a great extent fictional and contributed to the discussion and dissemination of the conflict in Britain via a selective imperial gaze.[11] According to Simon Popple, the demand for visual representations of a distant guerrilla war was answered by both the print media and by the new-born cinematograph where the standard context for the exhibition of films became increasingly patriotic.[12] The theme of atrocities committed by the enemy, which had already characterized reports of the Sepoy Mutiny and the Boxer Rebellion, was built upon off-screen narratives circulating in popular magazines, journals, and newspapers, which reported accounts of hideous outrages perpetrated by the Boers. The *Western Mail*, to give just one example, described how Boer forces in November 1899 raised white flags to indicate surrender only to fire a volley into their enemies at close range when the British advanced without firing.[13]

Echoing the success of Roger Fenton's pioneering photographs of the Crimean War, British filmmakers made of the Boer War the first military campaign to be extensively recorded by the moving picture camera. As Nicholas Daly has argued, the birth of cinema coincided with one of the first industrialized wars, and films of the Boer War were for their audiences an uncanny experience involving the sight of the recently dead, or those soon to die, in so-called "living-pictures". Films of the Boer War played a part in mediating the war to the British public and were very appealing for audiences in the music halls in which they were shown. Cinema, which was even referred to as "Boerograph" or the "Wargraph", brought the war home,

10 Chanan, 241.

11 See Popple, "Startling, realistic, pathetic: The Mitchell and Kenyon 'Boer War' Films", 150–157.

12 S. Popple, "But the Khaki-Covered Camera is the *Latest* Thing: the Boer War Cinema and Visual Culture in Britain" in *Young and Innocent? The Cinema in Britain 1896–1930*, ed. Andrew Higson (Exeter: University of Exeter Press, 2002), 13–27.

13 *Western Mail* (Cardiff, Wales), Wednesday, 8 November 1899; Issue 9504.

its events acting as a stimulus to the emergent film industry.[14] Nevertheless, films of the Boer War were mainly restricted to scenes shot on location of preparation for and the aftermath of battles, or of troops departing from or returning to Britain, rather than the battles themselves. This is the case, for example, of the films of City Imperial Volunteers returning to London and marching from Paddington Station to the Guildhall which I discussed in this book.

The Boer War clearly influenced the image of urban space emerging from early films of South African cities. Several actuality films made in 1900 by British filmmakers represented British troops and military leaders marching through the streets of Cape Town and Johannesburg. These films did not display the categories of the exotic and native life, but rather emphasized the military supremacy of the forces of the British Empire, projecting a powerful and imposing image of itself. Edgar Hyman and Joe Rosenthal filmed several military scenes for the Warwick Trading Company in Cape Town: *Arrival and Reception of Lord Roberts at Cape Town*, *Naval Brigade Through Main Street, Port Elizabeth Troops and Artillery to the Front*, *CIVs March on Johannesburg*, and *Hoisting the Union Jack at Johannesburg*. A similar scene, *New South Wales Lancers Landing at Cape Town*, was filmed by William Dickson for Biograph Syndicate. Sydney Melsom filmed *City Imperial Volunteers Leaving Wellington Barracks* for Paul's Theatrograph, and C. Rider Noble photographed for Gibbons' Bio Tableaux *C.I.V.s at Cape Town*, showing Sir Alfred Milner inspecting troops and the CIVs handing over ammunition. Edgar Hyman's *Arrival and Reception of Lord Roberts at Cape Town* is one of the few films to have survived. Filmed on 10 January 1900, it shows Lord Frederick Sleigh Roberts, the commander who later relieved Kimberley and advanced to Pretoria, disembarking from the RMS Dunottar Castle and inspecting the guard of honour, getting into his carriage and driving away. A small contingent of soldiers is also shown marching along the quayside. In films such as this, Cape Town, which was absorbed by the empire in 1806, appears as a city defined by the British

14 Nicholas Daly, *Literature, Technology, and Modernity, 1860–2000* (Cambridge: Cambridge University Press, 2004), 56–75.

military presence. British troops and leaders like Lord Roberts are here represented as the heroes of a military campaign run for the benefit of the empire and its people.

On the other hand, no film offers hints of the biggest scandal of the Boer War, the controversial concentration camps, and the scorched earth tactics employed by British troops. The war and the camps led the economist John A. Hobson to formulate his radical critique of imperialism and jingoism in 1901, cultural characteristics which appeared in the unconditional support for the British Army typical of early films of the Boer War.[15] Indeed, as Elizabeth Grottle Strebel argued, both actuality films and staged re-enactments of the war "served as propaganda through selection, omission, and emphasis."[16] In a fiction film like Cecil Hepworth's *The Conjuror and the Boer*, where a Boer is enveloped in a Union Jack, the British are always heroic and the Boer villains. In actuality films, we can see the troops but no trace of British concentration camps can be seen. Welfare activist Emily Hobhouse, at that time secretary of the women's branch of the South African Conciliation Committee, visited camps at Port Elizabeth, Norvalspont, Aliwal North, Springfontein, Kimberley and Orange River. She described the living conditions in the camps in her *Report of a Visit to the Camps of Women and Children in the Cape and Orange River Colonies*:

> It can never be wiped out of the memories of the people. It presses hardest on the children. They droop in the terrible heat, and with the insufficient unsuitable food; whatever you do, whatever the authorities do, and they are, I believe, doing their best with very limited means, it is all only a miserable patch on a great ill. Thousands, physically unfit, are placed in conditions of life which they have not strength to endure.[17]

15 On Hobson's experiences as a journalist in the Boer War, see: Paula M. Krebs, *Gender, Race, and the Writing of Empire: Public Discourse and the Boer War* (Cambridge: Cambridge University Press, 1999), 22–26.

16 Elizabeth Grottle Strebel, "Imperialist Iconography of Anglo-Boer War Film Footage", in *Film Before Griffith*, ed. John L. Fell (Berkeley and Los Angeles: University of California Press, 1983), 267.

17 Emily Hobhouse, *Report of a visit to the camps of women and children in the Cape and Orange River colonies* (London: Friars Printing Association, 1901), 16.

Hobhouse's description of such miserable scenes, with entire villages rooted up, did not fit the celebratory military narratives of early films, and remained absent from representation. Like images of the slums of London, Manchester or Glasgow and the famine in China, these subjects were omitted and early filmmakers put forward a view of the conflict based on hero-worship. Miserable and violent scenes such as those described by Hobhouse would have cast an unacceptable shadow over the actions of the British troops in South Africa.

Films of South African cities made before the war displayed a different portrait of everyday life. Most of these peace-time films of Cape Town and Johannesburg are now thought lost, but the titles reveal a number of panoramic actuality films and street scenes with no reference to the military events in South Africa. In 1898, just before the Boer War, at least three British film companies sent their cameramen to Cape Colony to film scenes of everyday life in the South African colonial cities. R.W. Paul distributed a series of films called *Life in Cape Colony*, which included a number of Cape Town views. William Dickson made for Biograph Syndicate *Panoramic View of Table Bay*, the natural inlet overlooked by Cape Town. In the same year, the Warwick Trading Company had sent Edgard Hyman to Cape Colony to film its cities and inhabitants. Hyman filmed the streets of Cape Town from a moving tram and a trolley car in films such as *Adderley Street, Cape Town*, *Tram Ride Through Cape Town*, *Through Cape Town on a Tram*, *Cape Town: Ride Through Adderley Street*, *Ride on a Trolley Car in Cape Town (Strand Road)*, *Woodstock to Cape Town by Tram*, *Cape Town to Seapoint by Tram*. Hyman also made two longer series of films called *Cape Town* and *Port Elizabeth*.[18] The third city filmed by Hyman was Johannesburg, which had developed into a city from a small village following the discovery of gold in the 1880s: *Johannesburg: a Tram Ride*

18 *Cape Town* included "Panorama of Cape Town", "Cape Town docks", "The Sea Front of Cape Town", "Coaling Vessels by Natives", "Panorama of Table Bay", "Panorama of Table Mountain", "Cape Town from Sea", and "Cape Mounted Police." *Port Elizabeth* included "Tram Ride up Russell Road", "Tram Ride Through Main Street", "Climbing Cape Road, Panoramic view of Durban", "Trough Cape Road, Port Elizabeth", and "Down Whites Road, Port Elizabeth."

in Commissioner Street, *Johannesburg Handicap*, *Johannesburg Between the Chains*, and *Johannesburg*, which included scenes of the cattle and general market in Pritchard Street, were all filmed in 1898.

Hyman's *Adderley Street, Cape Town* is the only film known to have survived. It was shot from a stationary camera position and it shows horse-drawn traffic and a number of pedestrians. An electric tram can also be seen. The first electric tram in Cape Colony trundled up Adderley Street in 1896, as a result of considerable government expenditure encouraging urban expansion. Hyman's films of that street and *Ride on a Trolley Car in Cape Town*, filmed in the nearby Strand Road, showed a Victorian city, a street scene which did not differ greatly from those taken in Britain in the same period. Adderley Street was the very centre of Cape Town, recently abandoned as a place of residence but in the process of becoming the commercial centre of the city. As Vivian Bickford-Smith puts it, in Adderley Street a great number of "shops, warehouses, banks and insurance offices stood as a monument to the God that had created them, merchant capital."[19] In the 1890s Cape Town experienced a period of prosperity characterized by huge government investments in improving Table Bay harbour, the docks and public transport. Films like *Adderley Street, Cape Town*, which records this era, were distributed in Britain and showed to British audiences a wealthy urban space which might have reminded them of local cityscapes because of its buildings and electrified public transport. Hyman's films of Cape Town seems to offer a British, or Western image of the city, lacking in "natives" or "native" themes (with the only exception being locals supplying vessels with coal). Films of Johannesburg and Port Elizabeth are now lost, but it is safe to suggest that they offered a similar urban image: Commissioner Street and Pritchard Street were the chief business streets in Johannesburg and also represented a settler society with strong bonds to Britain. The same can be said for "Down Whites Road, Port Elizabeth", from the series *Port Elizabeth*, which was filmed in one of

19 Vivian Bickford-Smith, *Ethnic Pride and Racial Prejudice in Victorian Cape Town: Group Identity and Social Practice, 1875–1902* (Cambridge and New York: Cambridge University Press, 1995), 14.

the city's main thoroughfares, with its 1892 neoclassical opera house where on July 9 1896, the first cinematograph show in Port Elizabeth was put on.

These films do not engage the categories of the picturesque and the exotic, and indeed Cape Colony did not play the same role in the Victorian imagination as East Asia and India. It was only ten years after the Boer War that we find a series of films which focused more conspicuously on exotic activities pursued by the "natives". These films were made by Frank Butcher and Emile Lauste for Empire in 1911, and included *Day in the Life of a Rickshaw Boy*, *Johannesburg: the Golden City*, and *Scenes In and Around Cape Town*. *Day in the Life of a Rickshaw Boy* was filmed in Durban, in the province of Natal, and has survived. It shows a rickshaw boy preparing himself and his vehicle for work: we can see him cleaning his rickshaw, decorating his uniform and preparing his head-dress. This is followed by scenes of the rickshaw runners' work in the streets of the city. The streets are not busy and neither cars, trams nor buses can be seen. The rickshaws, driven by "native" black people for white settlers, are the only means of transport on the streets. The cityscape of Durban here appears radically different from the images of Cape Town and Johannesburg and, as much as actuality films made in the Indian colonies, its narrative is inscribed in the imperial theme and the category of the urban exotic.

As we have seen in this section, the cinematic image of cities in the settler colony in South Africa rarely emphasized the exoticism of the territory and focused almost exclusively on the white non-indigenous population. British filmmakers in North Africa and the Middle East, in contrast, focused on the categories of the exotic.

Mediterranean cityscapes: From Egypt to the Holy Land

Edward Ziter has argued that the Islamic East exercised a hold on the British imagination equal to India. Its cultural and spatial distance from Britain, Ziter suggests, defined the area known today as the Middle East as a space of antiquity and outside modernity, an exotic world created in the Western imagination.[20] Kasbah, bedouins, houris, minarets, archaeological ruins, olive trees and carpets: the image of the Arab world had long been bound to a number of exotic views free from the features of modernity when British filmmakers began to film the old cities of Egypt, Algeria, Palestine and Morocco. They filmed street scenes and cityscapes in North African countries which were not under British control. For example, both Hepworth Manufacturing and the Warwick Trading Company's cameramen filmed a number of scenes in Morocco and Algeria.[21] In particular, early filmmakers focused on the Holy Land and Egypt.

Under Ottoman rule, from 1841 to 1917, the Holy Land already exercised a strong hold on the British imagination and, fifteen years before the British Mandate of Palestine was drafted in 1917, Palestinian cities were the subject of actuality films. In 1905, F. Ormiston-Smith filmed for Charles Urban a series called *Jaffa and Jerusalem*. The series is divided into two parts: "Jaffa and its Harbour" and "Jerusalem the Holy City." The first part is introduced by a view filmed from the front of a railway engine en route for Jaffa and passing two small stations in an arid Mediterranean landscape. The following scene shows boats in the harbour at Jaffa, the harbour crowded with fishermen unloading goods from small boats against the background of Jaffa's old buildings. Everyday activities take place on the quayside, where we can see a caravan of camels carrying crates of oranges and Arab men dressed in traditional clothing. The second part of the film opens with two panoramas of Jerusalem, including temples and mosques

20 Ziter, 11–13.

21 In 1901, Hepworth produced and filmed *The Solar Eclipse*, a series including "Street Scenes in Algiers", "Tangier, the Landing Stage", and "Tangier, the Market Place." In 1907, Hepworth's cameraman, H.M. Lomas, filmed *In and around Tangier* and street scenes in Mogador. In 1902, C. Rider Noble filmed for the Warwick Trading Company scenes in Morocco and Algiers: *The Bioscope in Algiers* and *The Uprising in Morocco*. The first series included "Panorama of Market Square at Biskra", "Panorama of Oldest Mosque in Algeria", "Scenes in Sidi Okba", "Circular Panorama of Market Square", "Scenes in Kairouan", "Snapshots of Algeria", "Panorama from train approaching El Kantara". The second series included "Street life in the Sacred City of Mequinel" and "Moorish street minstrel performing at Morocco."

dominating the cityscape. These are followed by scenes of the unloading of a camel, women and children walking by the city wall, and views of the Temple area and the Mosque of Omar. These images tell us a lot about local atmosphere and customs at this time but actuality films of Egypt are perhaps even more revealing.

Timothy Mitchell has investigated the ways in which the more exotic elements of the Egyptian city were experienced by Western visitors from a privileged point of view which allowed them to rediscover, and extended, what they already knew from photographs, drawings, books, and exhibitions.[22] Filmmakers travelled to North Africa and filmed Egyptian cities with a focus on the exotic elements of native streets, people and customs. In 1882, the British defeated the Egyptian Army at Tel El Kebir and took control of the country and *de facto* control of the strategic trade route of the Suez Canal.[23] This event marked the beginning of British military occupation of Egypt that lasted until 1936. The occupation of Egypt was due to the collapse of the Khedival *régime*, and continued because of French hostility provoked by the occupation itself.[24] Formally part of the Turkish Empire but operated as a British protectorate, Egypt was far from being the centre of the British Empire and was never perceived as a British colony, either in Egypt or in Britain.[25] In Victorian Britain, Egypt was rather seen as one of the most exotic places on the world map and, in particular, its ancient monuments were a source of constant interest. The country had long been described in popular books like Edward Lane's *An Account of the Manners and Customs of the Modern Egyptians* (1836). Egypt was a tourist destination for the wealthy and also the subject of many Victorian

22 Timothy Mitchell, *Colonising Egypt* (Berkeley: University of California Press, 1992) 34–62.

23 John S. Galbraith, and Afaf Lufti al-Sayyid-Marsot, "The British Occupation of Egypt: Another View", *International Journal of Middle East Studies*, vol. 9, no. 4 (1978): 471–488.

24 See Ronald Robinson, and John Gallagher (with Alice Denny), *Africa and the Victorians: the Official Mind of Imperialism* (London: Macmillan, 1981), 462–472.

25 See Lawrence James, *The Rise & Fall of the British Empire* (London: Abacus, 2007), 272–273.

travel books.[26] Ancient Egyptian statues and artefacts were displayed at the British Museum in London or, in the case of the famous Cleopatra's Needle, even on the Victoria Embankment. Throughout the nineteenth century, new technologies such as lithography and photography were employed to bring Egypt and the Middle East within reach of the imperial imagination of exotic territories. In the 1840s, the artist David Roberts travelled to Egypt and the Holy Land and sketched exotic monuments and ruins. His work was reproduced by Louis Haghe in *The Holy Land, Idumea, Arabia, Egypt and Nubia* (142–149), five volumes obtained through the technique known as "lithotint" which emulated the effect of watercolour washes.[27] Early British photographers travelled to Egypt to record its monuments and landmarks: in the 1850s Francis Frith used the "collodion" process to photograph monuments, people and cities in Egypt, Palestine and Syria. His concern with the representation of the most characteristic aspects of landscapes has been called topographic, and it laid further claim for photography in the advancement of science and education.[28] As Derek Gregory has argued, early photographs of Egypt such as those taken by Frith contributed to the formation of a colonial imaginary through the production of abstract spaces where the everyday life of natives was seen as a performance on a particular stage. Egypt was thus exoticized and rendered as a geographical space that could be fully known by the colonial gaze. This knowledge was obtained through the work of photographers drawn to Cairo because it was a fashionable destination for middle-class European travellers.[29] Popular exhibitions such as Burford's panorama of

26 See Martin R. Kalfatovic, *Nile Notes of a Howadji: A Bibliography of Travelers' Tales From Egypt, From the Earliest Time to 1918* (Lanham, MD: Scarecrow Press, 1997).

27 See James R. Ryan, "Images and Impressions: Printing, Reproduction and Photography", in *The Victorian Vision: Inventing New Britain*, ed. John M. MacKenzie (London: V & A, 2001), 221.

28 See Beaumont Newhall, *The History of Photography from 1839 to the Present* (New York: MoMA Press, 1982), 103–115.

29 Derek Gregory, "Emperors of the Gaze: Photographic Practices and Productions of Space in Egypt. 1839–1914", in *Picturing Place: Photography and the Geographical Imagination*, eds Joanna M. Schwartz and James R. Ryan (London and New York: I.B. Tauris, 2003), 195–225.

the Valley of the Nile in Leicester Square in 1848 – which was, according to *The Satirist*, "worthy of public patronage" – also contributed to the consolidation of the orientalist image of Egypt, and this image migrated to early films.[30] Early cameramen filmed Egypt with a similar topographical approach and created a number of views of its monuments, as well as its cities and inhabitants. As much as early photographs, actuality films contributed to the creation of a fully recognizable landscape in the colonial imaginary. Views of Egypt in early films were available to a much larger audience than that reached by lithographs and photographs, which were often only available to wealthy patrons.

Egypt was filmed as a landscape defined by its ancient history, which was embodied by the surviving monuments, and picturesque views of cities and the quotidian activities of their inhabitants. The exoticization of the country was thus achieved by pioneer filmmakers by focusing on themes related to Egypt's landmarks and its folklore. The first moving images of Egypt, which have not survived, were filmed as early as 1897 by Henry Short for Paul's Theatrograph. Short's series *Egypt* included views of the ancient temples of Egypt as well as urban scenes in Cairo of "natives" engaged in everyday activities, such as "Cairo Scene: Selling Water" and "Loading Camels in a Cairo Street". A few years later, in 1901, the Warwick Trading Company produced a number of actuality films of Egypt, in most cases filmed by C. Rider Noble and Joe Rosenthal. These included *Circular Panorama of Port Said*, *From Cairo to the Pyramids* and *Street Scene in Cairo*. *The Bioscope in Egypt: Cairo to Khartoum* was the longest series hitherto produced and included seventy-seven films representing both Egyptian urban life and the remains of the ancient Egyptian civilization.[31] The Warwick films are now lost and in order to find surviving early British films of Egypt we need to look at Hepworth's expedition in 1909. In that year, Hepworth Manufacturing produced a long series of films called *Scenes of Egyptian Life or Land of the Pharaohs*. It was filmed

30 *The Satirist* (London, England), 19 September 1847.

31 In 1910, the Warwick Trading Company filmed again in Egypt: *King and Queen Disembark Escorted by Lord Kitchener the Sidar*.

by Scott Brown, and included "native" urban scenes such as bazaars and merchants at Cairo and a camel-race. One film from this series has survived: *Moonlight Trip on the Nile*, a remarkable representation of the river bank, with its fields and palm trees, and a mud hut village with a mosque and twin minarets. Here is the picturesque aspect of life in Egypt which helped to create an exotic image of the country. Films like *Moonlight Trip on the Nile* mediated the perceived distance of cultural dislocation, and resulted in a visual explanation of colonial relations in subdued terrains of the imperial cultural system.

The relationship between knowledge and power in the Age of Empire is also revealed by *Egypt and Her Defenders*, a surviving film produced in 1914 by Kineto, Charles Urban's new company. *Egypt and Her Defenders*, which included sights of Egypt and a review of British troops, consolidates the role of Egypt in the imperial system by emphasizing the military part played by Egypt in the defence of the empire. The film confirms the exoticism of Egypt to the eyes of Western observers by displaying famous and picturesque monuments and places in Cairo. The film opens with scenes of the busy waterfront at Port Said with boats and ships. This is followed by street scenes in the crowded Arab quarters of Port Said, where "natives" are engaged in everyday activities such as the delivery of goods. The streets are not crowded and the "natives" seem to look with suspicion at the camera. In contrast with earlier films, *Egypt and Her Defenders* displays a clear educational intent as the following images of the famous sights of Egypt are introduced by intertitles giving historical information about the Great Sphinx of Giza and the pyramids. The second part of the film was shot in Cairo, introduced by a panorama of the old city and its monuments, focusing on the mosque of As-Sultan Hasan. This is followed by images of the procession of the Sacred Carpet from Mecca: the cavalry, a band, camels, crowds, infantry and a canopy carrying the carpet pass through a crowd-lined street in the city centre. The following scenes show a review of Egyptian cavalry, lancers and artillery by the Khedive, the ruler of Egypt and Sudan. This is followed by several monumental views of the Mosque of Muhammad Ali and the Cairo Citadel, the former seat of the government. The closing scenes of *Egypt and Her Defenders* show a crowd watching Lord Kitchener and another officer on horseback reviewing Egyptian troops. A text explains that these troops, reinforced by Indian soldiers, were about to fight against the Turks in the Great War (on 5 August 1914, Egypt was declared to be at war with the enemies of Britain). In *Egypt and Her Defenders*, themes such as everyday life in Egyptian cities and ancient landmarks are accompanied by an imperial narrative which emphasizes the British presence in Egypt and serves as straightforward propaganda in view of the Great War and the fight against the Turks. More broadly, films such as this contribute to the understanding of cultural relationships between the imperial centre and its colonial periphery, and were at the heart of an imperial system which defined the idea of home, Britain, and otherness, the colonies and the dominions.

In this chapter we have seen how, between the Boer War and the First World War, cinema played an important part in building an artificial image of Africa and the Middle East for Western audiences and helped to spread interest in military events in the African territories of the empire. The study of films of cities in early British films has revealed a radical difference between occupied territories in the north and settler societies in South Africa. On one hand, films of Cairo and Jerusalem reveal an old, traditional and exotic image of life in the city; on the other, cities such as Cape Town and Johannesburg emerge as urban developments designed and experienced in the image of British cities.

Travelogues and films of events like the Boxer Rebellion, the Delhi Durbar or the Boer War revealed the educational and informative nature of cinema, later fully achieved in the newsreels. The academic debate surrounding early cinema has so far failed to assess fully its ethnographic value, in particular in relation to the life of people in the city. The informative and yet subjective nature of early films is revealed in the very large number of lost and extant films discussed in this chapter. One of the most striking examples of the educational intent of early cinema is represented by the work of the Anglo-American producer and filmmaker Charles Urban. Under the name Charles Urban Trading Company, he produced a number of informative travelogues. By 1911, his films consistently pursued an explicit educational purpose, often emphasized by descriptive titles. For example, the street scenes in *In Quaint Peking*, which have been discussed earlier in this chapter, were introduced by the title: "The European quarter of

Peking, one of the oldest cities in the world, and capital of China, a republic of 400,000,000 people." An expression of nineteenth-century popular culture, the informative potential of films was explored from its very early days, when filmmakers like Urban illustrated the world as a fantasy of imperial prospect. During the First World War, Urban became chairman of the British Topical Committee for War Films, for which he made *Britain Prepared* (1915), a propaganda film displaying scenes of Britain's military readiness, and edited the propaganda documentary *The Battle of the Somme* (1916). Urban's earlier films of Delhi, Peking, Cape Town or Dublin – as much as those made by other pioneer filmmakers – were already revealing a view of the world according to the order expressed by the British Empire.

As Assenka Oksiloff has argued, early cinema provided a substitution of the real through a screen and revealed a segment of the world isolated in a representation where the image took the place of material reality: cinema thus transformed reality into a world of experience.[32] In films of cities in colonized territories, a moving image of a street replaces the real place and with it the idea itself of the city. This image thus builds in the Western imagination a fictional place, a cinematic idea of the East constructed on earlier expressions of visual culture and display. The modern city is defined by the exclusion of its opposite. In the British cinematic city cultural diversity and exoticism were relegated to exhibitions and parades. In the colonial city, they represented the everyday. The unchanging gaze of the middle-class filmmaker recorded the spectacle of the exotic in the form of cultural display represented by films, new agents of empire contributing to the administration of a project of world knowledge and exploration. The Victorian view of Asia and Africa was exemplified by the sculptural allegories of the Prince Albert Memorial in Kensington Gardens: the glorious histories of India or Egypt were celebrated in sculptures addressing their timeless traditions. Implicitly, they suggested that only as parts of the British Empire could their ancient prestige could be restored. Early films reveal a similar duplicity towards the colonial subject: a fascination with eastern crafts, history and tradition was accompanied by a well-rooted support for the imperial

32 Assenka Oksiloff, *Picturing the Primitive* (New York: Palgrave, 2001), 2–10.

project. Filmmakers' duplicity towards their subjects thus mirrored Britain's complex engagement with the eastern terrains of her empire.

At the time when the first films where made, the Boer War and the Irish problem were exacerbating Britain's military and political anxieties, by now bound to economic and social anxieties caused by the fear of a downturn in world trade. Nevertheless, Britain's imperial confidence was promoted at home in urban ceremonies, exhibitions and films; overseas the problems of the empire were relegated from the screen. Early non-fiction films presented another story, one of British might and imperial order.

Conclusions

This study of the cinematic city thus aims at exploring that system of urban thoroughfares defined by Walter Benjamin as "a vascular network of imagination."[33] The investigation of the city in film can map the tension between what is represented on film and what is left off-screen; it thus addresses the problem of making visible the history of cities represented through processes which filter and select aspects and spaces of urban life. This investigation has enabled me to map early non-fiction films and define the ideology which produced them. By focusing on ideology and urban cultures, it has also proposed a framework of analysis for the investigation of the role played by non-fiction cinema in the creation of geographical imagination.

Since the turn of the twentieth century, filmmakers have produced diverse, radical and personal views in films which have tried to undermine the association between film and dominant ideology. Filmmakers have narrated the city in a wide variety of non-fiction films which shaped the

33 Walter Benjamin, "Material for the exposé of 1935", in *The Arcades Project* (ed. Rolf Tiedemann, trans. Howard Eiland and Kevin McLaughlin (Cambridge, MA, and London: Belknap Press), 901.

ways in which urban space is perceived. In the 1920s, city symphonies such as Paul Strand's *Manhatta* (1921), Dziga Vertov's *The Man with the Movie Camera* (1927) and Walter Ruttmann's *Berlin: Symphony of a Great City* (1927) presented montage sequences of urban spaces as a source of spectacle in their own right. Alfred Hitchcock himself planned a city symphony called *London* or *Life of a City*, which unfortunately never saw the light of day.[34] From the mid 1950s, documentaries such as Karel Reisz's *We Are the Lambeth Boys*, Lorenza Mazzetti's *Together*, Lindsay Anderson's *Every Day Except Christmas*, Alain Tanner and Claude Goretta's *Nice Time* and Robert Vas' *Refugee England* revealed new ways of exploring urban life with their poetic authenticity in typical *Free Cinema* fashion.[35] On the other hand, the relationship between non-fiction films and propaganda has also evolved. The first official propaganda films made in Britain were distributed by the British Topical Committee for War Films and included Charles Urban's *Britain Prepared* (1915) and Geoffrey Malins and John McDowell's *The Battle of the Somme* (1916). The Empire Marketing Board was established in 1926 and its film unit, led by John Grierson, produced over one-hundred films to promote and support Britain's imperial adventure during the interwar years. The 1930s also saw documentaries such as *Housing Problems* (1935), directed by Arthur Elton and produced by The British Commercial Gas Association. This film revealed a cinematic way of tackling urban concerns, such as the conditions of life in the slums, which was more problematic and less celebratory than that emerging from earlier films. The Crown Film Unit produced a number of propaganda films during the war, including Humphrey Jennings's *London Can Take It* (1940), *Fires Were Started* (1943) and *Battle of London* (1941).

While cinema thus turned into a more complex and mature medium, imperial cities turned into post-imperial spaces. Early films of London or Liverpool embodied and dramatized the imperial theme as a collective

34 Steven Jacob, *The Wrong House: the Architecture of Alfred Hitchcock* (Rotterdam: 010 Publishers, 2007), 42.

35 See Sue Harper and Vincent Porter, *British Cinema of the 1950s: the Decline of Deference* (Oxford: Oxford University Press), 2003.

mirror of the society whose consensus their landmarks and monuments expressed. The same sites, when not wiped out by German bombs or urban redevelopment, have survived the decay of the cultural matrix of imperialism, and monuments such as the Nelson's Column and spaces like the Port of Liverpool now contain a memory of past imperial might. Films such as Humphrey Jennings's *Family Portrait*, which was released in 1951 as part of the celebration of the Festival of Britain, aimed at assuring people of Britain's place in a new world order which saw the power of Britain fading away. The Festival of Britain, which took place exactly one hundred years after the Great Exhibition of 1851 in the first Crystal Palace in Hyde Park, was itself the subject of several amateur films representing an event aimed at raising the nation's spirits after the war and years of austerity. The festival offered a memory of past exhibitions and past affluence to the people of Britain, and the films show an image of a city experiencing the dissolution of its empire. They also recorded a picture of Labour's 1945 unrealized dream after the Second World War of a socialist Commonwealth built on planning and risk-taking. This type of documentary film footage of the city is now associated with nostalgia and the memories of those who lived during those years. Terrence Davies has used such footage in his recent film *Of Time and the City* (2008), a cinematic love song to Liverpool and a reflection on the ways in which the city has changed and on the passing of time, where old footage from the 1950s is combined with newly filmed images of the city. Cinematic urban spaces have thus been depicted in a variety of ways, engaging personal and public history and memory of the city.

The in-depth investigation of the city in non-fiction films thus reveals both a way to understand urban life in a given context and a way to designate an interdisciplinary approach to the history of documentary film. This field of research originates in the understanding of non-fiction filmmaking as an ideal medium in the recording of urban life. It prompts an investigation of the ways in which non-fiction cinema has emancipated itself from a state of submission to the forces of nineteenth-century imperialism and capitalism, and later turned into a striving record of life in the city by merging and juxtaposing urban narratives of class, racial, and sexual diversity. Documentary films such as Horace Ové's *Reggae* (1971), which focused on black youths on their way to a reggae festival at Wembley Stadium, or Adam Clapham's *Consenting Adults* (1967), which tackled the lives of gay men and women in London just prior to the enactment of the 1967 Sexual Offences Bill, reveal different narratives of the city. Just as early film generated tailor-made narratives for discourses of imperialism, later non-fiction films revealed the city as a practiced place with contesting narratives converging on the same site. Non-fiction cinema has thus developed from an expression of the dominant narrative of its time to a diverse mosaic of urban experiences defined by class, race and gender.

Bibliography

Books

Altick, Richard. *The Shows of London*. London and Cambridge, MA: Harvard University Press, 1978.

Annan, Thomas. *Old Closes and Streets of Glasgow, 1868/1877*. Glasgow: City Improvement Trust, 1879.

Annan, Thomas. *Photographs of the Old Closes and Streets, a Series of Photogravures, 1868–1899*. Glasgow: James Maclehose, 1900.

Armes, Roy. *A Critical History of the British Cinema*. New York: Oxford University Press, 1978.

Ashby, Justine, and Andrew Higson, eds. *British Cinema, Past and Present*. London and New York: Routledge, 2000.

Ashcroft, Bill, Gareth Griffins and Helen Tiffin, eds. *Post-Colonial Studies: The Key Concepts*. London: Routledge, 2003.

Ashton, Thomas S. *Economic and Social Investigations in Manchester, 1833–1933: a Centenary History of the Manchester Statistical Society*. Hassocks: Harvester Press, 1977.

Archer, Mildred. *Visions of India: The Sketchbooks of William Simpson 1859–62*. Oxford: Phaidon, 1986.

Attridge, Derek, and Marjorie Howes, eds. *Semicolonial Joyce*. Cambridge: Cambridge University Press, 2000.

Attridge, Steve. *Nationalism, Imperialism and Identity in Late Victorian Culture: Civil and Military Worlds*. New York: Palgrave Macmillan, 2003.

Baedeker, Karl, ed. *London and Its Environs: An Handbook for Travellers*. Leipzig: Karl Baedeker, 1900.

Bailey, Peter. *Popular Culture and Performance in the Victorian City*. Cambridge and New York: Cambridge University Press, 1998.

Ballantyne, Tony. *Orientalism and Race: Aryanism in the British Empire*. New York and Houndmills, Hampshire: Palgrave, 2002.

Baratay, Eric, and Elisabeth Hardouin-Fugier. *Zoo: A History of Zoological Gardens in the West*, London: Reaktion Books, 2003.

Barber, Stephen. *Projected Cities: Cinema and Urban Space*. London: Reaktion Books, 2002.

Barnes, John. *The Beginning of Cinema in England, 1894–1901*. Newton Abbot: David and Charles, 1976.

Barnes, John. *The Beginning of Cinema in England, 1894–1901: Jubilee Year 1897*. London: Bishopsgate Press, 1983.

Barnes, John. *The Beginning of Cinema in England, 1894–1901: Filming the Boer War*. London: Bishopsgate Press, 1992.

Barringer, Timothy, and Tom Flynn, eds. *Colonialism and the Object: Empire, Material Culture and the Museum*. London and New York: Routledge, 1998.

Barringer, Timothy, Geoff Quilley and Douglas Fordham, eds. *Art and the British Empire*. Manchester and New York: Manchester University Press, 2007.

Barsan, Richard M. *Non-Fiction Film: a Critical History*. Bloomington, Indiana: Indiana University Press, 1992.

Bartholomew, John. *Atlas of the British Empire throughout the World*. London, 1868.

Beckson, Karl. *London in the 1890s: a Cultural History*. New York and London: W.W. Norton & Company, 1992.

Bell, Frederick. *Midnight Scenes in the Slums of New York*. New York: John Howitt and Son: 1881.

Benjamin, Walter. *The Arcades Project*. ed. Rolf Tiedemann, trans. Howard Eiland and Kevin McLaughlin. Cambridge, MA and London: Belknap Press, 1999.

Benjamin, Walter. *One-way Street and Other Writings*. London and New York: Verso, 1985 (*Einbahnstraße*, originally published in 1928 in Berlin).

Berman, Marshall. *All That is Solid Melts Into Air: The Experience of Modernity*. New York and London: Penguin, 1988 (first published in the US by Simon & Schuster, 1982).

Bernstein, Matthew, and Gaylyn Studlar, eds. *Visions of the East: Orientalism in Film*. New Brunswick, NJ: Rutgers University Press, 1997.

Betjeman, John. *First and Last Loves*. London: J. Murray, 1969.

Bhabha, Homi K. *The Location of Culture*. London and New York: Routledge, 1994.

Bickford-Smith, Vivian. *Ethnic Pride and Racial Prejudice in Victorian Cape Town: Group Identity and Social Practice, 1875–1902*. Cambridge and New York: Cambridge University Press, 1995.

Black, Barbara J. *On Exhibit: Victorians and Their Museums*. Charlottesville and London: University Press of Virginia, 2000.

Booth, Charles. *Life and Labour of People in London*. London and Edinburgh: Williams and Norgate, 1891.

Bordwell, David, and Kristin Thompson. *Film History: An Introduction*. New York: McGraw-Hill, 1994.

Boyer, Christine. *The City of Collective Memory: Its Historical Imagery and Architectural Entertainments*. London and Cambridge, MA: MIT Press, 1994 (Fifth Printing: 2001).
Bradley, Simon, and Nikolaus Pevsner. *London 1: The City of London*. London: Penguin, 1997.
Brady, Joe, and Anngret Simms, eds. *Dublin: Through Space and Time (c. 900–1900)*. Ann Arbor: University of Michigan Press, 2002.
Brantlinger, Patrick. *Rule of Darkness: British Literature and Imperialism, 1830–1914*. Ithaca, NY: Cornell University Press, 1988.
Briggs, Åsa. *Victorian Cities*. London: Penguin, 1990.
Brown, Richard, and Anthony Barry. *A Victorian Enterprise: The History of the British Mutoscope and Biograph Company*. London: Flicks Books, 1997.
Bruno, Giuliana. *Atlas of Emotions, Journeys in Art, Architecture, and Film*. London and New York: Verso, 2002.
Bruno, Gilulianа. *Streetwalking on a Ruined Map: Cultural Theory and the City Film of Elvira Notari*. Princeton, NJ: Princeton University Press, 1992.
Brunsdon, Charlotte. *London in Cinema: the Cinematic City Since 1945*. London: BFI Publishing, 2008.
Buckner, Phillip, ed. *Canada and the British Empire*. Oxford: Oxford University Press, 2008.
Burch, Noël. *Life to Those Shadows*. London: BFI Publishing, 1990.
Cannadine, David. *Ornamentalism: How the British Saw Their Empire*. London: Penguin Books, 2001.
Carroll, Clare, and Patricia King, eds. *Ireland and Postcolonial Theory*. Notre Dame, IN, and Cork, Ireland: University of Notre Dame Press and Cork University Press, 2003.
Çelik, Zeynep, Diane Favro, and Richard Ingersoll, eds. *Streets: Critical Perspectives on Public Space*. Berkeley, Los Angeles and London: University of California Press, 1994.
Chanan, Michael. *The Dream that Kicks: The Prehistory and Early Years of Cinema in Britain*. London and New York: Routledge, 1996 (original edition: 1980).
Chattopadhyay, Swati. *Representing Calcutta: Modernity, Nationalism, and the Colonial Uncanny*. London and New York: Routledge, 2006.
Cheng, Vincent J. *Joyce, Race, and Empire*. Cambridge: Cambridge University Press, 1995.
Chowdhry, Prem. *Colonial India and the Making of Empire Cinema: Image, Ideology and Identity*. Manchester and New York: Manchester University Press, 2000.
Christie, Ian, and Richard Taylor, eds. *The Film Factory: Russian and Soviet Cinema in Documents, 1896–1939*. Cambridge, MA: Harvard University Press, 1988.

Clark, Peter, ed. *The Cambridge Urban History of Britain*. Vol. II and III, Cambridge: Cambridge University Press, 2000.
Clarke, David B. *The Cinematic City*. London: Routledge, 1997.
Clifford, James. *The Predicament of Culture*. Cambridge, MA, and London: Harvard University Press, 1988.
Cochran, Sherman, ed. *Inventing Nanjing Road: Commercial Culture in Shanghai, 1900–1945*, Ithaca, NY: Cornell University Press, 2000.
Collingridge, Vanessa. *Boudica: The Life of Britain's Legendary Warrior Queen*. New York: Overlook Press, 2007.
Conder, Josiah. *The Modern Traveller. A Description, Geographical, Historical and Topographical of the Various Countries of the Globe*. Vol. III, London: Duncan, 1830.
Coombes, Annie. *Reinventing Africa: Museums, Material Culture and Popular Imagination in Late Victorian and Edwardian England*. New Haven and London: Yale University Press, 1984.
Corkin, Stanley, *Realism and the Birth of the Modern United States*. Athens and London: University of Georgia Press, 1996.
Cosgrove, Denis. *Apollo's Eye: a Cartographic Genealogy of the Earth in the Western Imagination*. Baltimore and London: The John Hopkins University Press, 2001.
Crossick, Geoffrey, ed. *The Lower Middle Class in Britain 1870–1914*. London: Croom Helm, 1977.
Crowley, John. *Atlas of Cork City*. Cork: Cork University Press, 2005.
Daiches, David. *Glasgow*. London: André Deutsch, 1977.
Daly, Nicholas. *Literature, Technology, and Modernity, 1860–2000*. Cambridge: Cambridge University Press, 2004.
Daniels, Stephen. *Fields of Vision: Landscape Imagery and National Identity in England and the United States*. Princeton, NJ: Princeton University Press, 1993.
Dauton, Martin, and Bernhard Rigier. *Meanings of Modernity: Britain from the Late-Victorian Era to World War II*. Oxford and New York: Berg, 2001.
Davis, Andrew. *Leisure, Gender and Poverty: Working Class Culture in Salford and Manchester, 1900–1939*. Buckingham: Open University Press, 1992.
Davis, Mike. *City of Quartz: Excavating the Future in Los Angeles*. London: Verso, 1990.
Davis, Mike. *Late Victorian Holocausts: El Niño Famines and the Making of the Third World*. London and New York: Verso, 2001.
Dehejia, Vidya. *India through the Lens: Photography 1840–1911*. Washington: Freer Gallery of Art and Arthur M. Sackler Gallery, 2001.
Dennis, Richard. *English Industrial Cities of the Nineteenth Century: a Social Geography*. London and Cambridge: Cambridge University Press, 1984.
Dixon, Wheeler W., ed. *Re-Viewing British Cinema, 1900–1992: Essays and Interviews*. New York: State University of New York Press, 1994.

Donald, James. *Imagining the Modern City*. London: Athlone Press, 1999.
Driver, Felix, ed. *Geography Militant: Cultures of Exploration and Empire*. Oxford and Malden, MA: Blackwell, 2001.
Driver, Felix, and David Gilbert, eds. *Imperial Cities: Landscape, Display and Identity*. Manchester and New York: Manchester University Press, 2003 (first published in 1999).
Driver, Felix, and Luciana Martins, eds, *Tropical Visions in an Age of Empire*. Chicago and London: Chicago University Press, 2005.
Edwards, Brian, and Paul Jenkins, eds, *Edinburgh: the Making of a Capital City*. Edinburgh: Edinburgh University Press, 2005.
Engels, Friedrich. *The Condition of the Working Class in England*. Oxford: Basil Blackwell, 1971.
Erdman, David V., ed. *The Complete Poetry and Prose of William Blake*. London: Random House, 1988.
Fell, John L., ed. *Film Before Griffith*. Berkeley and Los Angeles: University of California Press, 1983.
Field, John H. *Toward a Programme of Imperial Life: the British Empire at the Turn of the Century*. Westport Connecticut: Greenwood Press, 1982.
Foucault, Michel. *Discipline and Punish: The Birth of the Prison*. London: Allen Lane, 1977.
Foucault, Michel. *The Order of Things: an Archaeology of Human Sciences*. New York: Vintage, 1994.
Foucault, Michel. *Power/Knowledge: Selected Interviews and Writings, 1972–1977*. New York: Pantheon, 1981.
Foster, Robert F. *Modern Ireland: 1600–1972*. London: Penguin Books, 1989.
Fraser, W. Hamish. *The Coming of the Mass Market: 1850–1914*. London and Basingstoke: Macmillan Press, 1981.
Fraser, W. Hamish, and Irena Maver, eds. *Glasgow: 1830 to 1912*. Vol. II, Manchester and New York: Manchester University Press, 1996.
Freeman, Michael J., and Derek H. Aldcroft, eds. *Transport in Victorian Britain*. Manchester and New York: Manchester University Press, 1988.
Frykenberg, Robert E., ed. *Delhi Through the Ages: Essays in Urban History, Culture and Society*. Oxford and Delhi: Oxford University Press, 1986.
Fullerton, John, and Jan Olsson, eds. *Nordic Explorations: Films Before 1930*. Sydney, John Libbey & Company Pty Ltd, 1999.
Gibb, Andrew. *Glasgow: the Making of a City*. London: Croom Helm, 1983.
Gifford, Denis. *The British Film Catalogue: Fiction Film, 1888–1994*. London and Chicago: Fitzroy Dearborn, 2000 (3rd edn).

Gifford, Denis. *The British Film Catalogue: Non-Fiction Film, 1888–1994*. London and Chicago: Fitzroy Dearborn, 2001 (3rd edn).

Gilbert, Pamela K., ed. *Imagined Londons*. New York: State University of New York Press, 2002.

Gordon, George, ed. *Perspectives of the Scottish City*. Aberdeen: Aberdeen University Press, 1985.

Gray, Frank. *The Hove Pioneers and the Arrival of Cinema*. Brighton: University of Brighton, 1996.

Gray, John. *City in Revolt: James Larkin & the Belfast Dock Strike of 1907*. Belfast: Blackstaff, 1985.

Greenhalgh, Paul. *Ephemeral Vistas: History of the Expositions Universelles, Great Exhibitions and World's Fairs*. Manchester and New York: Manchester University Press, 1988.

Grewal, Inderpal. *Home and Harem: Nation, Gender, Empire, and Cultures of Travel*. Durham and London: Duke University Press, 1996.

Griffiths, Alison. *Wondrous Difference: Cinema, Anthropology and Turn-of-the-century Visual Culture*. New York: Columbia University Press, 2002.

Gunn, Simon. *The Public Culture of the Victorian Middle Class: Ritual and Authority and the English Industrial City, 1840–1914*. Manchester: Manchester University Press, 2000.

Gunn, Simon, and Robert J. Morris, eds. *Identities in Space: Contested Terrains in the Western City since 1850*. Aldershot, Burlington, Singapore and Sydney: Ashgate, 2001.

Gunning, Tom. *D.W. Griffith and the Origins of American Narrative Film: the Early years at Biograph*. Urbana: University of Illinois Press, 1991.

Gupta, Narayani. *Delhi Between Two Empires 1803–1931: Society, Government and Urban Growth*. Oxford and Delhi: Oxford University Press, 1981.

Gutsche, Thelma. *The History and Social Significance of Motion Pictures in South Africa, 1895–1940*. Cape Town: Howard Timms, 1972.

Hackforth-Jones, Jocelyn, and Mary Roberts, eds. *Edges of Empire: Orientalism and Visual Culture*. Malden, MA and Oxford: Blackwell, 2005.

Hall, Peter. *Cities in Civilization: Culture, Innovation, and Urban Order*. London: Phoenix Giant, 1999.

Hardy, Florence E., ed. *The Early Life of Thomas Hardy, 1840–1891*. London: Macmillan, 1928.

Harper, Sue, and Vincent Porter. *British Cinema of the 1950s: the Decline of Deference*. Oxford: Oxford University Press, 2003.

Harvey, David. *The Condition of Postmodernity: an Enquiry into the Origins of Cultural Change*. Oxford and New York: Basil Blackwell, 1989.

Hatch, Ernest F. *Far Eastern Impressions*. London: Huchinson, 1905.
Hepworth, Cecil. *Came the Dawn*. London: Phoenix House, 1951.
Herlihy, David. *Bicycle: The History*. New Haven: Yale University Press, 2004.
Herbert, Reginald. *Narrative of a Journey through the Upper Provinces of India, from Calcutta to Bombay, 1824–1825, with Notes upon Ceylon, and Account of a Journey to Madras and the Southern Provinces, 1826, and Letters written in India*. Vol. 1, London: Murray, 1828.
Higson, Andrew, ed. *Dissolving Views: Key Writings on British Cinema*. London: Cassell, 1996.
Higson, Andrew. *Waving the Flag: Constructing a National Cinema in Britain*. Oxford: Clarendon Press, 1995.
Higson, Andrew, ed. *Young and Innocent? The Cinema in Britain 1896–1930*. Exeter: University of Exeter Press, 2002.
Hill, John, and Kevin Rockett, eds. *Film History and National Cinema*. Dublin: Four Courts, 2005.
Hill, John, and Kevin Rockett, eds. *National Cinemas and World Cinema*. Dublin and Portland, Oregon: Four Courts, 2006.
Hobhouse, Emily. *Report of a visit to the camps of women and children in the Cape and Orange River colonies*. London: Friars Printing Association, 1901.
Hobsbawm, Eric. *The Age of Capital: Europe 1848–1875*. London: Weidenfeld and Nicolson, 1975.
Hobsbawm, Eric. *The Age of Empire: 1875–1914*. London: Abacus, 2006 (first edition published by Weidenfeld & Nicolson, 1987).
Hobsbawm, Eric. *The Age of Revolution: Europe 1789–1848*. London: Weidenfeld and Nicholson, 1962.
Hobsbawm, Eric. *Industry and Empire: From 1750 to Present Day*. London: Penguin, 1968.
Hobsbawm, Eric, and Terence Ranger, eds. *The Invention of Tradition*. Cambridge: Cambridge University Press, 1983.
Hobson, John A. *The Psychology of Jingoism*. London: Grant Richards, 1901.
Hodder, C. *Through London by Omnibus; a Chatty Guide to the Principal Sights of London: A Chatty Guide to the Principal Sights of London*. London: Hodder, 1898.
Hoffenberg, Peter. *An Empire on Display*. Berkeley: University of California Press, 2001.
Horrall, Andrew. *Popular Culture in London c. 1890–1918: The Transformation of Entertainment*. Manchester and New York: Manchester University Press, 2001.
Howe, Stephen. *Ireland and Empire: Colonial Legacies in Irish History and Culture*. Oxford: Oxford University Press, 2000.

Hunt, Tristram. *Building Jerusalem: The Rise and Fall of the Victorian City*. London: Orion Books, 2005 (first published by Weidenfeld & Nicolson, London, in 2004).

Hyde, Francis E. *Liverpool and the Mersey: The Development of a Port 1700–1970*. Newton Abbot: David & Charles, 1971.

Inden, Ronald D. *Imagining India*. Oxford: Basil Blackwell, 1990.

Jacob, Steven. *The Wrong House: The Architecture of Alfred Hitchcock*. Rotterdam: 010 Publishers, 2007.

James, Lawrence. *The Rise and Fall of the British Empire*. London: Abacus, 2007.

Jameson, Fredric. *The Political Unconscious: Narrative as a Socially Symbolic Act*. London and New York: Routledge, 1989.

Jameson, Fredric. *Postmodernism, or, the Cultural Logic of Late Capitalism*. London and New York: Verso, 1991.

Jeffery, Keith, ed. *An Irish Empire? Aspects of Ireland and the British Empire*. Manchester: Manchester University Press, 1996.

Joyce, James. *Dubliners*. London: Penguin Classics, 2000 (first published in 1914).

Joyce, James. *Portrait of the Artist as a Young Men*. London: Penguin Classics, 2000 (first published in 1916).

Joyce, James. *Ulysses*. London: Penguin Classics, 2000 (first published in 1922).

Joyce, Patrick. *The Rule of Freedom: Liberalism and the Modern City*. London and New York: Verso, 2003.

Judd, Denis. *Empire: the British Imperial Experience from 1765 to the Present*. London: HarperCollins, 1996.

Judd, Denis, and Keith Surridge, eds. *The Boer War*. London: John Murray, 2002.

Kalfatovic, Martin R. *Nile Notes of a Howadji: A Bibliography of Travelers' Tales From Egypt, From the Earliest Time to 1918*. Lanham, MD: Scarecrow Press, 1997.

Kearns, Gerard and Charles W.J. Withers, eds. *Urbanising Britain: Essays on Class and Community in the Nineteenth Century*. Cambridge and New York: Cambridge University Press, 1991.

Kern, Stephen. *The Culture of Time and Space: 1880–1918*. Cambridge, MA and London: Harvard University Press, 2003 (originally published in 1983).

Kidd, Alan. *Manchester*. Keele: Ryburn, 1996.

Kidd, Alan. *State, Society, and the Poor in Nineteenth-Century England*. Basingstoke: Macmillan; New York: St Martin's Press, 1999.

Kidd, Alan, and David Nicholls, eds. *Gender, Civic Culture, and Consumerism: Middle-class Identity in Britain, 1800–1940*. Manchester: Manchester University Press, 1999.

Kidd, Alan, and Kenneth W. Roberts, eds. *City, Class and Culture: Studies of Social Policy and Cultural Production in Victorian Manchester*. Manchester: Manchester University Press, 1985.

Kirby, Lynne. *Parallel Tracks: the Railroad and Silent Cinema*. Exeter: University of Exeter Press, 1997.

Koven, Seth. *Slumming: Sexual and Social Politics in Victorian London*. Princeton, NJ and Oxford: Princeton University Press, 2004.

Kracauer, Siegfried. *Nature of Film: The Redemption of Physical Reality*. London: Denis Dobson, 1961.

Kracauer, Siegfried. *Theory of Film: the Redemption of Physical Reality*. Princeton, NJ: Princeton University Press, 1997 (originally published in 1960, London and New York: Oxford University Press).

Krebs, Paula M. *Gender, Race, and the Writing of Empire: Public Discourse and the Boer War*. Cambridge: Cambridge University Press, 1999.

Landau, Paul S., ed. Images and Empire: Visuality *in Colonial and Postcolonial Africa*. Berkeley: University of California Press, 2002.

Lane, Tony. *Liverpool: Gateway of Empire*. London: Lawrence & Wishart, 1987.

Langton, John, and Robert J. Morris, eds. *Atlas of Industrializing Britain, 1780–1914*. London: Methuen, 1986.

Ledger, Sally, and Roger Luckhurst, eds. *The Fin de Siècle: a Reader in Cultural History c. 1880–1900*. Oxford and New York. Oxford University Press, 2000.

Leersen, Joep. *Remembrance and Imagination: Patterns in the Historical and Literary Representation of Ireland in the Nineteenth Century*. Cork and Notre Dame, IN: Cork University Press and University of Notre Dame Press, 1997.

Lefebvre, Henri. *Introduction to Modernity*. trans. by John Moore, London and New York: Verso, 1995 (first published as *Introdution à la modernité*, Paris: Les Editions de Minuit, 1962).

Lefebvre, Henri. *The Production of Space*, trans. by Donald Nicholson-Smith, Oxford: Blackwell, 1991 (first published as *La production de l'espace*, Paris: Anthropos, 1974).

Lefebvre, Martin, ed. *Landscape and Cinema*. New York and London: Routledge, 2006.

Levell, Nicky. *Oriental Visions: Exhibitions, Travel, and Collecting in the Victorian Age*. London: Horniman Museum and Gardens, 2000.

Lindsay, Maurice. *Portrait of Glasgow*. London: Robert Hale & Company, 1972.

Low, Rachael. *The History of the British Film: 1896–1906*. London: Gerges Allen & Unwin, 1948.

Low, Rachael. *The History of the British Film: 1906–1914*. London: Gerges Allen & Unwin, 1949.

Lowe, Donald. *History of Bourgeois Perception*. Chicago: University of Chicago Press, 1982.

Lowry, Donal, eds. *The South African War Reappraised*. Manchester: Manchester University Press, 1999.

Mace, Rodney. *Trafalgar Square: Emblem of Empire*. London: Lawrence & Wishart, 2005.

McDowell, John K., and John B.S. Gilfillan. *The People's History of Glasgow: An Encyclopedic Record of the City from the Prehistoric Period to the Present Day*. Glasgow and London: Hay Nisbet, 1899.

MacKenzie, Hugh. *The Third Statistical Account of Scotland: The City of Aberdeen*. Edinburgh: Oliver and Boyd, 1953.

MacKenzie, John M., ed. *Imperialism and Popular Culture*. Manchester and New York: Manchester University Press, 1986.

MacKenzie, John M. *Orientalism: History, Theory and the Arts*. Manchester: Manchester University Press, 1995.

MacKenzie, John M., ed. *Popular Imperialism and the Military: 1850–1950*. Manchester: Manchester University Press, 1992.

MacKenzie, John M. *Propaganda and Empire*. Manchester and New York: Manchester University Press, 1986.

MacKenzie, John M., ed. *The Victorian Vision: Inventing New Britain*. London: Victoria and Albert Museum Press, 2001.

Makepeace, Chris, ed. *Manchester and Salford in the 1890s: A Collection of Photographs by Samuel Coulthurst*. Hendon Mill: Hendon Publishing, 1986.

Manning, Paul. *News and News Sources: a Critical Introduction*. New York: SAGE, 2001.

Marcus, Laura, and Lynda Nead, ed. *The Actuality of Walter Benjamin*. London: Lawrence and Wishart, 1998.

Marriott, John. *The Other Empire: Metropolis, India and Progress in the Colonial Imagination*. Manchester and New York: Manchester University Press, 2003.

Mayhew, Henry. *London Labour and the London Poor: The Condition and Earnings of those that will work, cannot work, and will not work*. London: C. Griffin, 1861.

Mayhew, Henry. *London's Underworld*. London: Spring Books, 1950.

Mazumdar, Ranjani. *Bombay Cinema: An Archive of the City*. Minneapolis: University Of Minnesota Press, 2007.

McCaffrey, John F. *Scotland in the Nineteenth Century*. Houndmills, Basingstoke: Macmillan, 1998.

McKernan, Luke. *Topical Budget: The Great British News Film*. London: BFI Publishing, 1991.

Messinger, Gary S. *Manchester in the Victorian Age: the Half-known City*. Manchester: Manchester University Press, 1985.

Michie, Helena, and Ronald Thomas. *Nineteenth-Century Geographies: the Transformation of Space from the Victorian Age to the American Century*. New Brunswick, NJ, and London: Rutgers University Press, 2003.

Midwinter, Eric. *Old Liverpool*. Newton Abbot: David & Charles, 1971.
Mitchell, Timothy. *Colonising Egypt*. Berkeley: University of California Press, 1992.
Mitchell, William J.T., eds. *Landscape and Power*. Chicago and London: University of Chicago Press, 2002 (first published in 1994).
Morris, Robert J., ed. *Atlas of Industrializing Britain, 1780–1914*. London: Methuen, 1986, pp. 94–105.
Morris, Robert J., ed. *Class, Power and Social Structure in British Nineteenth-Century Towns*. Leicester: Leicester University Press, 1986.
Morris, Robert J., and Richard Rodger, eds. *The Victorian City: a Reader in British Urban History: 1820–1914*. London and New York: Longman, 1993.
Mullaney, Steven. *The Place of the Stage: License, Play and Power in Renaissance England*, Chicago: University of Chicago Press, 1988.
Munro, Ian. *The Figure of the Crowd in Early Modern London: the City and its Double*. New York: Palgrave Macmillan, 2005.
Musser, Charles. *The Emergence of Cinema: the American Screen to 1907*. University of California Press: Berkeley and Los Angeles, 1994.
Nead, Lynda. *The Haunted Gallery Painting, Photography and Film around 1900*. London: Yale University Press, 2007.
Nead, Lynda. *Victoria Babylon: People, Streets and Images in Nineteenth-Century London*. New Heaven: Yale University Press, 2000.
Neal, Frank. *Sectarian Violence: the Liverpool Experience, 1819–1914: an Aspect of Anglo-Irish History*. Manchester: Manchester University Press, 1988.
Newhall, Beaumont. *The History of Photography from 1839 to the Present*. New York: MoMA Press, 1982.
Nichols, Bill. *Ideology and the Image*. Bloomington: Indiana University Press, 1981.
Oakley, Charles A. *The Second City: The Story of Glasgow*. Glasgow and London: Blackie, 1946.
Oakley, Charles A. *Where We Came In: Seventy Years of the British Film Industry*. London: George Allen and Unwin, 1964.
O'Brien, Harvey. *The Real Ireland: the Evolution of Ireland in Documentary Film*. Manchester University Press, 2004.
Oksiloff, Assenka. *Picturing the Primitive*. New York: Palgrave, 2001.
Orwell, George. *Nineteen Eighty-Four*. London: Penguin, 2000.
Pakenham, Thomas. *The Boer War*. London: Macdonald, 1979.
Palliser, David M., Peter Clark and Martin Daunton, eds. *The Cambridge Urban History of Britain, 1840–1950*. Vol. 3. Cambridge: Cambridge University Press, 2000.
Palmberg, Mai. *Encounter Images in the Meetings Between Africa and Europe*. Uppsala, Sweden: Nordic Africa Institute, 2001.
Panofsky, Erwin. *Meaning in the Visual Arts*. Harmondsworth: Penguin, 1970.

Panofsky, Erwin. *Studies in Iconography: Humanistic Themes in the Art of the Renaissance.* New York and Oxford: Oxford University Press, 1939.
Penz, François, and Thomas, Maureen, eds. *Cinema and Architecture.* London: BFI Publishing, 1997.
Perry, George. *The Great British Picture Show.* London: Pavilion Books, 1974 (second impression 1985).
Picard, Liza. *Victorian London: the Life of a City 1840–1870.* London: Weidenfeld & Nicolson, 2006.
Plomer, William, ed., *Kilvert's Diary, 1870–1879.* Harmondsworth, London: Penguin, 1982.
Poe, Edgar Allan. The Fall of the House of Usher and Other Writings. London: Penguin, 2003.
Porter, Andrew, ed. *The Oxford History of the British Empire: The Nineteenth Century,* Vol. III. Oxford and New York: Oxford University Press, 1999.
Porter, Bernard. *The Absent-Minded Imperialist: Empire, Society, and Culture in Britain.* Oxford: Oxford University Press, 2004.
Porter, Laraine, and Bryony Dixon, eds. *Picture Perfect: Landscape, Place and Travel in British Cinema before 1930.* Exeter: the Exeter Press, 2007.
Porter, Roy. *London: a Social History.* London: Penguin Books, 2000.
Prakash, Gyan. *After Colonialism: Imperial Histories and Postcolonial Displacements.* Princeton, NJ: Princeton University Press, 1995.
Prakash, Gyan. *Another Reason: Science and the Imagination of Modern India.* Princeton, NJ: Princeton University Press, 1999.
Preston, Diana. *The Boxer Rebellion: The Dramatic Story of China's War on Foreigners that Shook the World in the Summer of 1900.* New York: Penguin, 2001.
Price, Richard. *An Imperial War and the British Working Class: Working-Class Attitudes and Reactions to the Boer War, 1899–1902.* London: Routledge and Kegan Paul, 1972.
Pulling, Christopher. *They were singing and what they sang about.* London: George G. Harrap, 1952.
Reeves, Nicholas. *The Power of Film Propaganda: Myth or Reality?* London and New York: Cassell, 1999.
Rhodes, John David. *Stupendous, Miserable City: Pasolini's Rome.* Minneapolis: University Of Minnesota Press, 2007.
Richards, Thomas. *The Commodity Culture of Victorian England.* Chicago and London: Stanford University Press, 1990.
Richards, Thomas. *The Imperial Archive: Knowledge and the Fantasy of Empire.* London and New York: Verso, 1993.

Robinson, Ronald, and John Gallagher (with Alice Denny), eds. *Africa and the Victorians: the Official Mind of Imperialism*. London: Macmillan, 1981.
Rockett, Kevin. *Cinema and Ireland*. London: Croom Helm, 1987.
Rockett, Kevin, Luke Gibbons and John Hill, eds. *Cinema and Ireland*. London: Routledge, 1987.
Rodger, Richard. *The Transformation of Edinburgh: Land, Property and Trust in the Nineteenth Century*. Cambridge: Cambridge University Press, 2001.
Ross, Robert, and George J. Telkamp. *Colonial Cities: Essays on Urbanism in a Colonial Context*. Dordrecht, Netherlands: Martinus Nijhoff, 1985.
Rousseau, George S., and Roy Porter, eds. *Exoticism in the Enlightenment*. New York and Manchester: New York University Press, 1990.
Ryan, James R. *Picturing Empire: Photography and the Visualization of the British Empire*. London: Reaktion Books, 1997.
Sadoul, George. *Histoire Générale du cinema: 2. Les pionniers du cinéma 1897–1909*. Paris: Denoël, 1948.
Said, Edward. *Culture and Imperialism*. London: Vintage, 1994.
Said, Edward. *Orientalism: Western Conceptions of the Orient*. London: Penguin Books, 1995 (first published by Routledge and Kegan Paul, 1978).
Saint, Andrew, ed. *Politic and People of London: the London County Council 1889–1965*. London and Roncervete: Hambledon Press, 1989.
Samuel, Raphael, ed. *Patriotism: The Making and Unmaking of British National Identity (Volume II: Minorities and Outsiders)*. London and New York: Routledge, 1989.
Sanders, James. *Celluloid Skyline: New York and the Movies*. New York: Alfred A. Knopf, 2001.
Schneer, Jonathan. *London 1900: The Imperial Metropolis*. New Haven, CT and London: Yale University Press, 1999.
Schivelbusch, Wolfgang. *The Railway Journey: The Industrialization of Time and Space in the 19th Century*. Leamington, Hamburg and New York: Berg, 1986.
Schwartz, Joanna M., and James R. Ryan, eds. *Picturing Place: Photography and the Geographical Imagination*. London and New York: I.B. Tauris, 2003.
Schwartz, Vanessa. *Spectacular Realities: Early Mass Culture in* Fin-de-Siècle *Paris*. Berkeley, Los Angels and London: University of California Press, 1998.
Schwartz, Vanessa, and Jeannene M. Przyblyski, eds. *Nineteenth-Century Visual Culture*. New York and London: Routledge, 2004.
Seeley, John Robert. *The Expansion of England*. Cambridge: Cambridge University Press, 2005 (first published in 1883).
Sennett, Richard. *The Fall of the Public Man*. London: Faber and Faber, 1986.
Sennett, Richard. *Flesh and Stone: the Body of the City in Western Civilization*. London and Boston, MA: Faber and Faber, 1994.

Shaw, George Bernard. *Fabianism and the Empire: A Manifesto by the Fabian Society.* London: Fabian Society, 1900.
Shiel, Mark. *The Real Los Angeles: Hollywood, Cinema, and the City of Angels.* London and Chicago: Reaktion Books, 2009.
Shiel, Mark, and Tony Fitzmaurice, eds. *Cinema and the City: Film and Urban Societies in a Global Context.* Oxford and New York: Blackwell, 2001.
Shiel, Mark, and Tony Fitzmaurice, eds. *Screening the City.* London and New York: Verso, 2003.
Shipman, David. *The Story of Cinema: From the Beginning to Gone with the Wind.* London: Hodder and Stoughton, 1982.
Shohat, Ella. *Unthinking Eurocentrism: Multiculturalism and the Media.* London and New York: Routledge, 1994.
Smith, John S., and David Stevenson. *Aberdeen in the Nineteenth Century: The Making of the Modern City.* Aberdeen: Aberdeen University Press, 1988.
Smith Headingly, Adolphe, and John Thompson, *Street Life in London.* London: Sampson Low, Marston, Searle, & Rivington, 1877.
Soja, Edward W. *Postmodern Geographies: The Reassertion of Space in Critical Social Theory.* London and New York: Verso, 1989.
Sprinker, Michael, ed. *Edward Said: a Critical Reader.* Oxford, 1992.
Stoker, Bram. *Dracula.* London: Penguin, 1993.
Stollery, Martin, *Alternative Empires: European Modernist Cinemas and Cultures of Imperialism.* Exeter: University of Exeter Press, 2000.
Storch, Robert D., ed. *Popular Culture and Custom in Nineteenth-Century England.* London and Canberra: Croom Helm, 1982.
Sutcliffe, Anthony. *Metropolis 1810 –1940.* London: Mansell Publishing, 1984.
Swinton, William. *Atlas.* Cambridge, MA: Ivison, Blakeman and Taylor, 1875.
Thomas, Nicholas. *Colonialism's Culture: Anthropology, Travel and Government.* Oxford and Cambridge: Polity Press, 1994.
Tomaselli, Keyan G. *Appropriating Images: The Semiotics of Visual Representation.* Charlottesville: University of Virginia Press, 1994.
Toulmin, Vanessa. *Electric Edwardians: The Story of the Mitchell & Kenyon Collection.* London: BFI Publishing, 2006.
Toulmin, Vanessa, Simon Popple, and Patrick Russell. *The Lost World of Mitchell and Kenyon: Edwardian Britain on Film.* London: British Film Institute, 2004.
Townshend, Charles. *Ireland: the 20th Century.* London: Arnold; New York: Oxford University Press, 1999.
Vidler, Anthony. *The Architectural Uncanny: Essays in the Modern Unhomely.* Cambridge, MA: MIT Press, 1992.

Volwahsen, Andreas. *Imperial Delhi: The British Capital of the Indian Empire.* Munich, Berlin, London and New York: Prestel, 2002.
Walkowitz, Judith R. *Prostitution and Victorian Society: Women, Class and the State.* Cambridge: Cambridge University Press, 1980.
Walton, John K. *The British Seaside: Holidays and Resorts in the Twentieth Century.* Manchester: Manchester University Press, 2000.
Ward, Stephen V. *Selling Places: the Marketing and Promotion of Towns and Cities, 1850–2000.* London: Taylor and Francis, 1998.
Weller, Edward. *Portable Atlas of the British Empire.* London, 1875.
White, Jerry. *London in the Nineteenth Century.* London: Jonathan Cape, 2007.
Williams, Raymond. *The Country and the City.* London: Chatto & Windus, 1973.
Ziter, Edward. *The Orient on the Victorian Stage.* Cambridge and New York: Cambridge University Press, 2003.

Chapters in books

Adams, Annemarie. "The Healthy Victorian City: the Old London Street at the International Health Exhibition of 1884." In *Streets: Critical Perspectives on Public Space*, edited by Zeynep Çelik, Diane Favro, and Richard Ingersoll, 203–212. Berkeley, Los Angeles and London: University of California Press, 1994.
Arscott, Caroline. "The representation of the city in the visual arts", in *The Cambridge Urban History of Britain, vol. 3: 1840–1950*, edited by David M. Palliser, Peter Clark, Martin Daunton, 811–831. Cambridge: Cambridge University Press, 2000.
Barker, Theodore C. "Urban Transport", in *Transport in Victorian Britain*, edited by Michael J. Freeman and Derek H. Aldcroft, 134–170. Manchester and New York: Manchester University Press, 1988.
Barringer, Tim. "The South Kensington Museum and the colonial project", in *Colonialism and the Object: Empire, Material Culture and the Museum*, edited by Tim Barringer and Tom Flynn, 11–27. London and New York: Routledge, 1998.
Benjamin, Walter. "Material for the exposé of 1935", in *The Arcades Project*, edited by Rolf Tiedemann (trans. Howard Eiland and Kevin McLaughlin), 899–918. Cambridge, Mass., and London: Belknap Press, 1999.
Bottomore, Stephen, "Frederic Villiers: War Correspondent", in *Re-Viewing British Cinema, 1900–1992: Essays and Interviews*, edited by Wheeler W. Dixon, 11–23. New York: State University of New York Press, 1994.

Cannadine, David. "The Context, Performance and Meaning of Ritual: the British Monarchy and the *Invention of Tradition*, c. 1820–1877", in *The Invention of Tradition*, edited by Eric Hobsbawm and Terrence Ranger, 101–164. New York and Cambridge: Cambridge University Press, 2000.

Canosa, Michele. "Batman: la linea gotica" in *Desideri informa di nuvole, cinema e fumetto*, edited by Michele Canosa and Enrico Fornaioli, 185–201. Campanotto Editore: Prato, 1996.

Cohn, Bernard S. "Representing Authority in Victorian India", in *The Invention of Tradition*, edited by Eric Hobsbawm and Terence Ranger, 165–209. Cambridge: Cambridge University Press, 1983.

Condon, Denis. "Spleen of a cabinet minister at work: exhibiting X-ray and the cinematograph in Ireland, 1896", in *Film History and National Cinema*, edited by John Hill and Kevin Rockett, 69–78. Dublin: Four Courts, 2005.

Costa, Antonio. "Landscape and Archive: Trips around the World as Early Film Topic (1896–1914)", in *Landscape and Cinema*, edited by Martin Lefebvre, 245–266. New York and London: Routledge, 2006.

Cowman, Krista. "The battle of the boulevards: class, gender and the purpose of public space in later Victorian Liverpool", in *Identities in Space: Contested Terrains in the Western City since 1850*, edited by Simon Gunn and Robert J. Morris, 152–164. Aldershot, Burlington, Singapore and Sydney: Ashgate, 2001.

Dennis, Richard. "Modern London", in *The Cambridge Urban History of Britain, vol. 3: 1840–1950*, edited by David M. Palliser, Peter Clark, Martin Daunton, 95–132. Cambridge: Cambridge University Press, 2000.

Dicks, Brian. "Choice and constraint: further perspectives on socio-residential segregation in nineteenth-century Glasgow with particular reference to its West End", in *Perspectives of the Scottish City*, edited by George Gordon, 91–124. Aberdeen: Aberdeen University Press, 1985.

Driver, Felix. "Geographical Knowledge, Exploration and Empire", in *Geography Militant: Cultures of Exploration and Empire*, edited by Felix Driver, 1–48. Oxford and Malden, MA: Blackwell Publishers, 2001.

Driver, Felix, and David Gilbert. "Imperial Cities: Overlapping Territories intertwined Histories", in *Imperial Cities: Landscape, Display and Identity*, edited by Felix Driver and David Gilbert, pp. 1–17. Manchester and New York: Manchester University Press, 2003.

Driver, Felix, and Luciana Martins. "Views and Visions of the Tropical World", in *Tropical Visions in an Age of Empire*, edited by Felix Driver and Luciana Martins, 3–20. Chicago and London: Chicago University Press, 2005.

Duffy, Edna. "Disappearing Dublin: *Ulysses*, postcoloniality and the politics of space", in *Semicolonial Joyce*, edited by Derek Attridge and Marjorie Howes, 37–57. Cambridge: Cambridge University Press, 2000.

Feldman, David. "Jews in London: 1880–1914", in *Patriotism: the Making and Unmaking of British National Identity (Volume II: Minorities and Outsiders)*, edited by Raphael Samuel, 207–230. London and New York: Routledge, 1989.

Fitzpatrick, David. "Ireland and the Empire", in *The Oxford History of the British Empire: the Nineteenth Century*, vol. III, edited by Andrew Porter, 494–521. Oxford and New York: Oxford University Press, 1999.

Fraser, W. Hamish. "Municipal Socialism and Social Policy", in *The Victorian City: a Reader in British Urban History: 1820–1914*, edited by Robert J. Morris and Richard Rodger, 258–280. London and New York: Longman, 1993.

Frykenberg, Robert E. "The Coronation Durbar of 1911: Some Implications", in *Delhi Through the Ages: Essays in Urban History, Culture and Society*, edited by Robert E. Frykenberg, 369–390. Oxford and Delhi: Oxford University Press, 1986.

Fullerton, John. "Seeing the World with Different Eyes, or Seeing Differently", in *Nordic Explorations: Films Before 1930*, edited by John Fullerton and Jan Olsson, 163–186. Sydney: John Libbey & Company Pty Ltd, 1999.

Gilbert, David, and Fiona Henderson. "London and the tourist imagination" in *Imagined Londons*, edited by Pamela K. Gilbert, 121–136. New York: State University of New York Press, 2002.

Gilbert, Pamela K. "Introduction: Imagining Londons", in *Imagined Londons*, edited by Pamela K. Gilbert, 1–10. New York: State University of New York Press, 2002.

Gillespie, James. "Municipalism, Monopoly, Management: the Demise of Socialism in One Country: 1918–1933", in *Politic and People of London: the London County Council 1889–1965*, edited by Andrew Saint, 103–133. London and Roncervete: Hambledon Press, 1989.

Gregory, Derek. "Emperors of the Gaze: Photographic Practices and Productions of Space in Egypt. 1839–1914", in *Picturing Place: Photography and the Geographical Imagination*, edited by Joanna M. Schwartz, and James R. Ryan, 195–225. London and New York: I.B. Tauris, 2003.

Grottle Strebel, Elizabeth. "Imperialist Iconography of Anglo-Boer War Film Footage", in *Film Before Griffith*, edited by John L. Fell, 264–271. Berkeley and Los Angeles: University of California Press, 1983.

Gunn, Simon. "The middle class, modernity and the provincial city: Manchester, c. 1840–80", in *Gender, Civic Culture, and Consumerism: Middle-class Identity in Britain, 1800–1940*, edited by Alan Kidd and David Nicholls, 112–127. Manchester: Manchester University Press, 1999.

Gunning, Tom. "Cinema of Attractions: Early Cinema, Its Spectator and the Avant-Garde", in *Early Cinema: Space Frame Narrative*, edited by Thomas Elsaesser. 56–62. London: BFI Publishing, 1990.

Gunning, Tom. "Images of the City in Early Cinema", unpublished paper presented at the Getty Centre, *Cinecity Film and Perceptions. of Urban Space*, Santa Monica, California, 23 March 1994, in H. Weihsmann, "The City in Twilight: Charting the Genre of the City Film 1900–1930", 9.

Gunning, Tom. "Pictures of Crowd Splendour: the Mitchell and Kenyon Factory Gate Films", in *The Lost World of Mitchell and Kenyon*, edited by Vanessa Toulmin, Simon Popple and Patrick Russell, 49–58. London: BFI Publishing, 2004.

Hansen, Peter. "Modern Mountains: The Performative Consciousness of Modernity in Britain, 1870–1940", in *Meanings of Modernity: Britain from the Late-Victorian Era to World War II*, edited by Martin Dauton and Bernhard Rigier, 185–202. Oxford and New York: Berg, 2001.

Hassam, Andrew. "Portable iron structures and uncertain colonial spaces at the Sydenham Crystal Palace", in *Imperial Cities, Landscape, Display and Identity*, edited by Felix Driver, and David Gilbert, 174–193. Manchester and New York: Manchester University Press, 2003 (first published in 1999).

Hume, David. "Empire Day in Ireland, 1896–1962", in *An Irish Empire? Aspects of Ireland and the British Empire*, edited by Keith Jeffery, 149–168. Manchester: Manchester University Press, 1996.

Jackson, Alvin. "Ireland, the Union, and the Empire, 1800–1960", in *Ireland and the British Empire*, edited by Kevin Kenny, 123–153. Oxford: Oxford University Press, 2004.

Jackson, Gordon. "Sea trade", in *Atlas of Industrializing Britain, 1780–1914*, edited by John Langton and Robert J. Morris, 94–105. London: Methuen, 1986.

Jeffery, Keith. "Introduction", in *An Irish Empire? Aspects of Ireland and the British Empire*, edited by Keith Jeffery, 1–24. Manchester: Manchester University Press, 1996.

Jeffery, Keith. "The Irish military tradition and the British empire", in *An Irish Empire? Aspects of Ireland and the British Empire*, edited by Keith Jeffery, 94–122. Manchester: Manchester University Press, 1996.

Keiller, Patrick. "The City of the Future", in *Picture Perfect: Landscape, Place and Travel in British Cinema before 1930*, edited by Laraine Porter and Bryony Dixon, 104–112. Exeter: Exeter University Press, 2007.

Keiller, Patrick. "Tram Rides and Other Virtual Landscapes", in *The Lost World of Mitchell and Kenyon: Edwardian Britain on Film*, edited by Toulmin, Vanessa, Simon Popple and Patrick Russell, 191–200. London: BFI Publishing, 2004.

Keiller, Patrick. "Urban space and early cinema", in *Cities in Transition: the Moving Image and the Modern Metropolis*, edited by Andrew Weber and Emma Wilson, 29–39. London: Wallflower Press, 2008.

Kidd, Alan. "Outcast Manchester: Voluntary Charity, Poor Relief and the Casual Poor 1860–1905", in *Class and Culture: Studies of Social Policy and Cultural Production in Victorian Manchester*, edited by Alan Kidd and Kenneth W. Roberts *City*, 48–73. Manchester: Manchester University Press, 1985.

King, Anthony D. "Colonial Cities: Global Pivots of Change", in *Colonial Cities: Essays on Urbanism in a Colonial Context*, edited by Robert Ross, and George J. Telkamp, 7–32. Dordrecht, Netherlands: Martinus Nijhoff, 1985.

Landau, Paul S. "Empires of the Visual: Photography and Colonial Administration in Africa", in Images and Empire: Visuality *in Colonial and Postcolonial Africa*, edited by Paul S. Landau, 140–171. Berkeley: University of California Press, 2002.

Le Bon, Gustave. "The Mind of Crowds", in *The Fin de Siècle: a Reader in Cultural History c. 1880–1900*, edited by Sally Ledger and Roger Luckhurst, 55–61. Oxford and New York: Oxford University Press, 2000 (first published by T. Fisher Unwin, London, in 1896).

McCole, Niamh. "The cinematograph in provincial Ireland 1896–1906: exhibition and reception", in *National Cinemas and World Cinema*, edited by John Hill and Kevin Rockett, 31–41. Dublin and Portland, Oregon: Four Courts, 2006.

MacKenzie, John M. "Empire and Metropolitan Cultures", in *The Oxford History of the British Empire: The Nineteenth Century*, edited by Andrew Porter, 270–293. Oxford and New York: Oxford University Press, 1999.

MacKenzie, John M. "'The Second City of the Empire': Glasgow – imperial municipality", *Imperial Cities: Landscape, Display and Identity*, edited by Felix Driver and David Gilbert, 215–237. Manchester and New York: Manchester University Press, 2003 (first published in 1999).

Marshall, Peter J. "Taming the exotic: the British and India in the seventeenth and eighteenth centuries", in *Exoticism in the Enlightenment*, edited by George S. Rousseau and Roy Porter, 46–65. New York and Manchester: New York University Press, 1990.

Meath, "British youth and the Empire", in *King and Country*, London, December 1902.

Mellor, Robert. "Aberdeen: the Great Century", in *Aberdeen in the Nineteenth Century: the Making of the Modern City*, edited by John S. Smith and David Stevenson, 1–18. Aberdeen: Aberdeen University Press, 1988.

Mitchell, William J.T. "Imperial Landscape", in *Landscape and Power*, edited by William J.T. Mitchell, 5–34. Chicago and London: University of Chicago Press, 2002 (first published in 1994).

Monks, Robert. "The Irish Films in the Mitchell and Kenyon Collection", in *The Lost World of Mitchell and Kenyon: Edwardian Britain on Film*, edited by Vanessa Toulmin, Simon Popple, and Patrick Russell, 93–102. London: British Film Institute, 2004.

Morgan, Nick F. "Building the city", in *Glasgow: 1830 to 1912*, Vol. II, edited by W. Hamish Fraser and Irena Maver, 8–55. Manchester and New York: Manchester University Press, 1996.

Morris, Robert J. "Urbanization", in *Atlas of Industrializing Britain, 1780–1914*, edited by John Langton and Robert J. Morris, 164–179. London: Methuen, 1986.

Palmer, Sarah. "Ports", in *The Cambridge Urban History of Britain*, Vol. II and III, edited by Peter Clark, 133–150. Cambridge: Cambridge University Press, 2000.

Osterhammel, Jürgen. "Britain and China", in *The Oxford History of the British Empire: The Nineteenth Century*, vol. III, edited by Andrew Porter, 146–169. Oxford and New York: Oxford University Press, 1999.

Pagani, Catherine. "Chinese material culture and British perception of China in the mid-nineteenth century", in *Colonialism and the Object: Empire, Material Culture and the Museum*, edited by Tim Barringer, and Tom Flynn, 28–40. London and New York: Routledge, 1998.

Popple, Simon. "But the Khaki-Covered Camera is the *Latest* Thing: the Boer War Cinema and Visual Culture in Britain", in *Young and Innocent? The Cinema in Britain 1896–1930*, edited by Andrew Higson, 13–27. Exeter: University of Exeter Press, 2002.

Popple, Simon. "Startling, realistic, pathetic: The Mitchell and Kenyon 'Boer War' Films", in *The Lost World of Mitchell and Kenyon: Edwardian Britain on Film*, edited by Vanessa Toulmin, Simon Popple and Patrick Russell, 150–157. London: BFI Publishing, 2004.

Prescott, Andrew. "'We had fine banners': Street Processions in the Mitchell and Kenyon Films" in *The Lost World of Mitchell and Kenyon*, edited by Vanessa Toulmin, Simon Popple and Patrick Russell, 125–136. London: BFI Publishing, 2004.

Price, Richard N. "Society, Status and Jingoism: the Social Roots of Lower Middle Class Patriotism, 1870–1900", in *The Lower Middle Class in Britain 1870–1914*, edited by Geoffrey Crossick, 89–112. London: Croom Helm, 1977.

Richards, Jeffrey. "Ireland, the Empire and film", in *An Irish Empire? Aspects of Ireland and the British Empire*, edited by Keith Jeffery, 25–56. Manchester: Manchester University Press, 1996.

Rousseau, George S., and Roy Porter. "Introduction", in *Exoticism in the Enlightenment*, edited by George S. Rousseau and Roy Porter, 46–65. New York and Manchester: New York University Press, 1990.

Ryan, Deborah S. "Staging the Imperial City: The Pageant of London, 1911", in *Imperial Cities, Landscape, Display and Identity*, edited by Felix Driver and David

Gilbert, 117–135. Manchester and New York: Manchester University Press, 2003 (first published in 1999).
Ryan, James R. "Images and Impressions: Printing, Reproduction and Photography", in *The Victorian Vision: Inventing New Britain*, edited by John M. MacKenzie, 215–39. London: Victoria and Albert Museum Press, 2001.
Said, Edward. "Orientalism reconsidered", in *Literature, Politics and Theory*, edited by Francis Becker 210–229. London: Methuen, 1987.
Said, Edward. "Reflections on Ireland and Postcolonialism", in *Ireland and Postcolonial Theory*, Clare Carroll and Patricia King, 177–185. Notre Dame, Indiana, and Cork, Ireland: University of Notre Dame Press and Cork University Press, 2003.
Schivelbusch, Wolfgang. "Panoramic Travel", in *The Nineteenth-Century Visual Culture*, edited by Vanessa R. Schwartz and Jeannene M. Przyblyski, 92–99. New York and London: Routledge, 2004.
Schmiechen, James. "Glasgow of the Imagination: Architecture, Townscape and Society", in *Glasgow: 1830 to 1912*, Vol. II, edited by W. Hamish Fraser and Irena Maver, 486–518. Manchester and New York: Manchester University Press, 1996.
Schwartz, Joanna M., and James R. Ryan. "Introduction: Picturing Place: Photography and the Geographical Imagination", in *Picturing Place: Photography and the Geographical Imagination*, edited by Joanna M. Schwartz and James R. Ryan, 1–18. London and New York: I.B. Tauris, 2003.
Schwartz, Vanessa, and Jeannene M. Przyblyski. "Visual Culture's History", in *The Nineteenth-Century Visual Culture*, edited by Vanessa Schwartz, and Jeannene M. Przyblyski, 3–14. New York and London: Routledge, 2004.
Shiel, Mark. "Cinema and the City in History and Theory", in *Cinema and the City: Film and Urban Societies in a Global Context*, edited by Mark Shiel and Tony Fitzmaurice, Oxford and New York: Blackwell Publishing, 2001.
Smith, Joan. "Class, skill and sectarianism in Glasgow and Liverpool, 1880–1914", in *Class, Power and Social Structure in British Nineteenth-Century Towns*, edited by Robert J. Morris, 157–204. Leicester: Leicester University Press, 1986.
Storch, Robert D. "Introduction: Persistence and Change in Nineteenth-Century Popular Culture", in *Popular Culture and Custom in Nineteenth-Century England*, edited by Robert D. Storch, 1–19. London and Canberra: Croom Helm, 1982.
Summerfield, Penny. "Patriotism and Empire: Music-Hall Entertainment 1870–1914", in *Imperialism and Popular Culture*, edited by John M. MacKenzie, 17–48. Manchester and New York: Manchester University Press, 1986.
Thompson, Francis M. "The Rise of Suburbia", in *The Victorian City: a Reader in British Urban History: 1820–1914*, edited by Robert J. Morris and Robert Rodger, 149–180. London and New York: Longman, 1993.

Thompson, John H. "Canada and the 'Third British Empire', 1901–1939", in *Canada and the British Empire*, edited by Phillip Buckner, 87–106. Oxford: Oxford University Press, 2008.

Vadillo, Ana. "Poetics on the Line: the Effect of Mass Transport in Urban Culture", in *Nineteenth-Century Geographies: the Transformation of Space from the Victorian Age to the American Century*, edited by Helena Michie and Ronald Thomas, 237–250. New Brunswick, NJ, and London: Rutgers University Press, 2003.

Weihsmann, Helmut. "The City in Twilight: Charting the Genre of the City Film 1900–1930", in *Cinema and Architecture*, edited by François Penz and Maureen Thomas, 8–27. London: BFI Publishing, 1997.

Yearsley, Ian. "On the Move in the Streets: Transport Films and the Mitchell and Kenyon Collection", in *The Lost World of Mitchell and Kenyon: Edwardian Britain on Film*, edited by Vanessa Toulmin, Simon Popple and Patrick Russell, 181–190. London: BFI Publishing, 2004.

Articles in journals

Benjamin, Walter. "Paris: Capital of the Nineteenth Century." *Perspecta* 12 (1969): 165–172.

Bertozzi, Marco. "Visualising the Past: the Italian City in Early Cinema." *Film History* 12, no. 3 (2000): 322–329.

Bottomore, Stephen. "An Amazing Quarter Mile of Moving Gols, Gems and Genealogy: Filming India's 1902/03 Delhi Durbar." *Historical Journal of Film, Radio and Television* 15, no. 4 (1995): 495–515.

Bottomore, Stephen. "Have You Seen the Gaekwar Bob?: Filming the 1911 Delhi Durbar." *Historical Journal of Film, Radio and Television* 17, no. 3 (1997): 309–345.

Bottomore, Stephen. "Rediscovering early non-fiction film." *Film History* 13, no. 2 (2001): 160–173.

Breckenridge, Carol. "The Aesthetics and Politics of Colonial Collecting: India at World Fairs." *Comparative Studies in Society and History* 31, no. 2 (1989): 195–216.

Brown, Simon. "Early Cinema in Britain and the Smoking Concert Film." *Early Popular Visual Culture* 3, no. 2 (2005): 165–178.

Bruno, Giuliana. "Site-seeing: Architecture and the Moving Image." *Wide Angle* 19, no. 4 (1997): 8–24.

Burrows, Jon. "Penny Pleasures: Film Exhibition in London During the Nickelodeon Era, 1906–1914." *Film History* 16, no. 1 (2004): 60–91.
Burrows, Jon. "Penny Pleasures II: Indecency, Anarchy and Junk Film in London's 'Nickelodeons', 1906–1914." *Film History* 16, no. 2 (2004): 172–197.
Burrows, Jon. "*A Vague Chinese Quarter Elsewhere*: Limehouse in the Cinema, 1914–36." *Journal of British Cinema and Television* 6, no 2 (2009): 282–301.
Burton, Antoinette. "London and Paris through Indian spectacles. Making a spectacle of Empire: Indian travellers in fin-de-siecle London." *History Workshop Journal* 42 (1996): 127–146.
Crook, William. "Nudity in India in Custom and Ritual." *The Journal of the Royal Anthropological Institute of Great Britain and Ireland* 49 (1919): 237–251.
Driver, Felix, and David Gilbert. "Capital and Empire: Geographies of Imperial London." *GeoJournal* 51, no. 1 (2000): 23–32.
Driver, Felix, and David Gilbert. "Heart of empire? Landscape, Space and Performance in Imperial London." *Environment and Planning D: Society and Space* 16, no. 1 (1998): 11–28.
Galbraith, John S., and Al-Sayyid-Marsot. "The British Occupation of Egypt: Another View." *International Journal of Middle East Studies* 9, no. 4 (1978): 471–488.
Gilbert, David. "London in all its glory – or how to enjoy London: Guidebook representations of Imperial London." *Journal of Historical Geography* 25, no. 3 (1999): 279–297.
Gunning, Tom. "From the Kaleidoscope to the X-Ray: Urban Spectatorship, Poe, Benjamin, and *Traffic in Souls*." *Wide Angle* 19, no. 4 (1997): 25–61.
Gunning, Tom. "The World as Object Lesson: Cinema Audiences, Visual Culture, and the St Louis World's Fair, 1904." *Film History* 6, no. 4 (1994): 422–44.
Hays, Michael. "Empire: Class, Culture, and the Popular Theatre in the Nineteenth Century." *Theatre Journal* 47, no. 1 (1995): 65–82.
Koeck, Richard. "Liverpool in Film: J.A.L. Promio's Cinematic Urban Space", *Early Popular Visual Culture* 7, no. 1 (2009): 63–81.
Nead, Lynda. "Animating the Everyday: London on Camera c.1900." *Journal of British Studies* 43, no. 1 (2004): 65–90.
Paul, Robert W. "Before 1910: Kinematograph Experiences." *Proceedings of the British Kinematograph Society*, no. 38 (1936): 4–5.
Raack, Richard C. "Historiography as Cinematography: A Prolegomenon to Film Work for Historians." *Journal of Contemporary History* 18, no. 3 (1983): 411–438.
Sadoul, George (translated by Y. Templin). "Early Film Production in England: the Origin of Montage, Close-Ups, and Chase Sequence." *Hollywood Quaterly* 1, no. 3 (1946): 249–259.

Schwarze, Tracey T. "Silencing Stephen: Colonial Pathologies in Victorian Dublin." *Twentieth Century Literature* 43, no. 3 (1997): 243–263.
Thompson, Andrew S. "The Language of Imperialism and the Meanings of Empire: Imperial discourse in British politics, 1895–1914." *The Journal of British Studies* 36, no. 2 (1997): 147–177.
Trevithick, Alan. "Some Structural and Sequential Aspects of the British Imperial Assemblages at Delhi, 1911." *Modern Studies* 24, no. 3 (1990): 561–578.
Vidler, Anthony. "The Explosion of Space: Architecture and the Filmic Imaginary." *Assemblage: A Critical Journal of Architecture and Design Culture* 21 (1993): 45–59.

Newspapers

The Builder, vol. 10, 1852.
Daily Mail, 3 February 1900.
Daily News, 16 March 1896.
The Era, 31 October 1896.
The Era, 10 October 1896.
The Era, 22 November 1840.
The Graphic, 2 May 1895.
The Halfpenny Marvel, 6 October 1900.
The Illustrated London News, No. 2646, Vol. XCVII (4 January 1890).
The Illustrated London News, No. 2964, Vol. CVIII (8 February 1896).
The Illustrated London News, No. 2973 vol. CVIII (11 April 1896).
The Illustrated London News, No. 2982, Vol. VIII (13 June 1896).
The Illustrated London News, No. 2985, Vol. CIX (4 July 1896).
The Illustrated London News, No. 3034, Vol. CX (12 June 1897).
The Illustrated London News, No. 3036, Vol. CX (26 June 1897).
The Illustrated London News, No. 3324, Vol. CXXII (3 January 1903).
The Illustrated London News, No. 3326, Vol. CXXII (17 January 1903).
The Illustrated London News, No. 3798, Vol. CXL (3 February 1912).
The Illustrated Weekly News, 12 October 1862.
Anon., "Trafalgar Day in Liverpool: Celebration of the Anniversary – A Patriotic Demonstration", *Liverpool Mercury*, 23 October 1899.
Liverpool Mercury 24 November 1900.
Anon., "Big Ben most iconic London film location", *Metro*, 21 October 2007: 8.

The Morning Chronicle, 8 April 1859.
Anon. (1901), "A commercial wonderland: Liverpool and its docks", *The New Penny Magazine*, no. 109, vol. IX: 177–184.
Gorky, Maxim, "The Lumière Cinematograph" in *Nizhnegorodskii Listok*, vol. 4, no. 7 (1896).
The Photogram, vol. 4, no. 43, July 1897.
The Satirist; or, the Censor of the Times (London, England), 27 April, 1845.
The Satirist; or, the Censor of the Times (London, England), 19 September 1847.
Anon., "Music and 'Effects' in Cinematography", *The Showman*, September 6, 1901: 574–75.
Sligo Champion, 17 January 1903.
The Times, Wednesday, 31 March 1897.
Anon., "Plague at Glasgow", *The Times*, Friday, 31 August 1900: 6.
Anon., "The outbreak of plague at Glasgow", *The Times*, Saturday, 22 September 1900: 9.
The Times, Monday, 21 January 1901.
The Times, Saturday, 15 February 1902.
The Times, Wednesday, 9 April 1913.
Trewman's Exeter Flying Post or Plymouth and Cornish Advertiser, 13 August 1857.
Western Mail, 8 November 1899.

Filmography

This filmography includes over five hundred films of cities classified in eight sections according to the place where they were filmed. Many company catalogues from the period 1895–1914 have not survived as they were not required by law to be submitted to official copyright deposits. This filmography includes nevertheless both extant and some lost films, and is based on my research at the British Film Institute and on the cataloguing work carried out by Denis Gifford.[1] Surviving films are marked with an asterisk (*). Throughout this book, film and series titles are in italics and scenes are in quotation marks.

England

London

At the Zoological Gardens (British Pathé, 1909)
Around the Paddington Engine Yard (Charles Urban Trading Company, 1904)
Band Marching Down a Street (Northern Photographic Works, Birt Acres, 1896) *
Bateau à vapeur sur la Tamise (Lumière Brothers, Alexandre Promio, 1896) *
Bateaux sur le lac de St James Park (Lumière Brothers, Alexandre Promio, 1896) *
Balloon Ascent at Crystal Palace (Interchangeable, 1898)
Bicycle Race at the Crystal Palace (George A. Smith, 1900)
Blackfriars Bridge (Paul's Animatograph Works, Robert W. Paul, 1896) *
Busy London – Traffic Passing in Front of the Bank of England and Mansion House (Walturdaw Company, 1903) *

1 Denis Gifford, *The British Film Catalogue: Volume 2, Non-Fiction Film, 1888–1994* (3rd edn), London and Chicago: Fitzroy Dearborn, 2001.

Carousel: Hampstead (1900) *
The City of London (Hepworth Manufacturing Company, Gaston Quiribet, 1909)
City of Westminster (Hepworth Manufacturing Company, Gaston Quiribet, 1909) *
Coaching in Hyde Park (Warwick Trading Company, Will Barker, 1909)
Corner of Barnet Fair (Northern Photographic Works, Birt Acres, 1896) *
Coronation Through London (James Williamson, 1902)
Covent Garden (J.B. McDowell, B&C, 1912)
Crowded Streets in London (Esmé Collings, 1896)
A Crowd in the City (Paul's Animatograph Works, Robert W. Paul, 1896)
Cycling in Hyde Park (Northern Photographic Works, Birt Acres, 1896)
The Crystal Palace and a Glimpse of the Zoo (New Agency, 1911)
Crystal Palace Panoramas (Charles Urban Trading Company, H.M. Lomas, 1903)
Danse Javanaise (Lumière Brothers, Alexandre Promio, 1896) *
Danseuses des rues (Lumière Brothers, Alexandre Promio, 1896) *
A Day in London (Gaumont, 1912)
Demolition of Old Railway Station, Croydon (Bender & Co, Adolphe Langfier, 1897)
Dèpart d'un bateau sur la Tamise (Lumière Brothers, Alexandre Promio, 1896) *
Disappointed London (Hepwix, Cecil Hepworth, 1902) *
Distinguished Guests Leaving the Power House (British Mutoscope and Biograph Company, William Kennedy-Laurie Dickson, 1901)
Elephants at the Zoo (Mutoscope and Biograph, William Dickson, 1898)
Empire Day (Warwick Trading Company, 1911)
Empire Theatre, Leicester Square (Lumière Brothers, Alexandre Promio, 1896)
Entree de Hyde Park (Lumière Brothers, Alexandre Promio, 1896) *
Evolution of London Traffic or the Passing of the Horse at the Mansion House (Hubsch, 1914)
Explosion of a Motor Car (Hepwix, Cecil Hepworth, 1901)
Feeding the Tigers (Warwick Trading Company, 1899) *
Finsbury Park Station (Northern Photographic Works, Birt Acres, 1896)
The First Trams Leaving Shepherd's Bush for Southall (British Mutoscope and Biograph Company, William Kennedy-Laurie Dickson, 1901)
Fountains (Paul's Animatograph Works, Robert W. Paul, 1897)
Funeral of Mr Gladstone (George A. Smith, 1898)
Gladstone's Funeral (Paul's Animatograph, Robert W. Paul, 1898)
Gladstone's Funeral (Warwick Trading Company, Cecil Hepworth, 1898)
Great Northern Mail Train Catching Post Bags (Northern Photographic Works, Birt Acres, 1897)
The Great Thames Pageant (Warwick Trading Company, Will Barker, 1909)
The Great Thames Pageant (Charles Urban Trading Company, 1909)

Head of the Colonial Procession Passing Westminster (Paul's Animatograph Works, 1897) *
Her Majesty the Queen Arriving at South Kensington on the Occasion of the Laying of the Foundation Stone of The Victoria And Albert Museum (Warwick Trading Company, 1900) *
Heroes of Ladysmith Marching Through London (Warwick Trading Company, 1900)
Historic Spots of Old London (Walter Tyler, 1911)
Horse Drawn Single Storey Tram in a London Street (Paul's Animatograph Works, Robert W. Paul, 1898)
Horse Drawn Traffic in Euston Road (Charles Goodwin Norton, 1898) *
Horse Drawn Traffic in Seven Sisters Road (Charles Goodwin Norton, 1898) *
Horse Drawn Traffic Viewed from Elevated Position (Charles Goodwin Norton, 1897) *
Horse Parade at Crystal Palace (Warwick Trading Company, 1899)
Hyde Park (William Friese-Greene 1889)
Hyde Park (Lumière Brothers, Alexandre Promio, 1896) *
Hyde Park Bicycling Scenes (Paul's Animatograph Works, Robert W. Paul, 1896) *
Hyde Park Corner (Chard's Vitagraph, 1897)
Hyde Park to the Bank in Three Minutes (Warwick Trading Company, 1899)
Houses of Parliament (Paul's Animatograph Works, Robert W. Paul, 1896) *
Houses of Parliament (George A. Smith, 1897)
International Ciclying Race at Crystal Palace (George A. Smith, Charles Urban Trading Company, 1906)
Irish Mail – L & N W Railway – Taking up Water at Full Speed (British Mutoscope and Biograph Company, 1898) *
Jongler Javanais (Lumière Brothers, Alexandre Promio, 1896) *
Lanchester car test on steps at Crystal Palace (1902) *
The Lane on Sunday Morning (Biograph Syndicate, William Dickson, 1899)
Laying of The Foundation Stone of the Victoria And Albert Museum (Biograph Syndicate, 1900) *Leaving Victoria Station* (Hepwix, Cecil Hepworth, 1901)
Liverpool Street Station Exterior (British Gaumont, 1906)
Liverpool Street Station Interior (British Gaumont, 1906)
Living London (Charles Urban Trading Company, 1906)[2] *

2 The series included: "Navvies on the Lumber Barges", "Flower Market at Covent Garden", "Covent Garden Market", "Sunday Morning at Petticoat Lane", "The London Coster at Work", "The London Coster at Play", "Costers' Cake Walk at Crystal Palace", "Old Father Thames", "Panorama of the Thames", "Life on the Upper Thames", "Hungerford and Charing Cross Bridges", "Tower Bridge – Raising Same", and "Bank of England and Mansion House".

Lions (Lumière Brothers, Alexandre Promio, 1896) *
London (J.W. Rowe's Pictorialograph, 1897)[3]
London (British Pathé, 1911)
London and its Life (Charles Urban Trading Company, 1910)[4]
London & North Western Railway Scotch Express (Charles Goodwin Norton, 1899) *
London as the King Sees It (Warwick Trading Company, Will Baker, 1905,)
The London Express (Paul's Animatograph Works, Robert W. Paul, 1899)
London: Factory Girls on Strike in Camden Town (1911)
London Fire Brigade at Hyde Park (Charles Urban Trading Company, 1906)
London Markets (Charles Urban Trading Company, 1910)[5]
London's New Museum (Charles Urban Trading Company, 1909)
London of Today (Gaumont, 1908)
London's Traffic at the Mansion House (Mutoscope and Biograph, William Dickson, 1898) *
London Street Scenes (Warwick Trading Company, 1899)[6]
London: The Coal Strike March (1912)
London: Zoological Gardens (Chard & Co, 1897)[7]
Looping the Loop (Charles Urban Trading Company, H.M. Lomas, 1903)
The Lord Mayor's Show (Chard's Vitagraph, 1897)
The Lord Mayor's Show (Gaumont, 1909)
Lord Roberts' Arrival in London (Warwick Trading Company, 1900)
Lutteurs Javanais (Lumière Brothers, Alexandre Promio, 1896) *
Marble Arch (Lumière Brothers, Alexandre Promio, 1896)
Marble Arch (Chard's Vitagraph, 1897)

3 "The Mansion House", "Cheapside", "Ludgate Circus", "Oxford Circus" and "Rotten Row."

4 "Social and Fashionable London", "Official London", "Political, Legal and Journalistic London", "Financial and Municipal London" and "Historic London and the Ghetto".

5 "Smithfield", Billingsgate" and "Covent Garden."

6 The series included: "Regent Street from Piccadilly", "Hyde Park Corner Looking West", "The Strand", "Charing Cross Station", "Westminster Clock Tower and St Margaret's", "Piccadilly Looking East", "Piccadilly from Green Park", "Piccadilly Circus", "Waterloo Place", "Trafalgar Square", "Pall Mall Looking East", "Whitehall", "The Strand Looking East", "The Strand Looking West", "The Law Courts – Strand", "Cheapside and Queen Victoria Street", "The Bank of England", "Traffic on London Bridge", "Traffic over Tower Bridge", "Tower Bridge", "Threadneedle and Princes Street", and "Fleet Street Looking East".

7 "Feeding the Pelicans" and "Polar Bears".

Moonlight on the Thames (Hepworth Manufacturing Company, Gaston Quiribet, 1909)
Motor Car Climbing Contest at the Crystal Palace (1902)
Motor Car Race in Regent's Park (William Dickson, 1900)
Nègres dansant dans la rue (Lumière Brothers, Alexandre Promio, 1896) *
The Nelson Column (William Dickson for Mutoscope and Biograph, 1897)
Nelson's Monument (George A. Smith, 1897)
Old London Street Scenes (1900) *
On a Runnaway Motor Car Through Piccadilly Circus (Paul's Animatograph, Works Robert W. Paul, 1899)
One Sort of Sunday Crowd at Hyde Park (Warwick Trading Company, 1900)
Opening of the New Electric Tram Route (British Mutoscope and Biograph Company, 1903)[8]
Opening of the South African Exhibition (Warwick Trading Company 1907)
Opening Ceremony at International Universal Exhibition (Chard's Vitagraph, 1898)
The Pageant of the Empire (Will Barker, 1911)
The Pageant of London (John Y. Brown, London Cinematograph Co., 1911)
Panorama at Ealing Showing Lord Rothschild Declaring Line Open (British Mutoscope and Biograph Company, William Kennedy-Laurie Dickson, 1901)
Panorama du Palais de Westminster pris de la Tamise (Lumière Brothers, Alexandre Promio, 1896) *
Panorama of Ealing from a Moving Tram (British Mutoscope and Biograph Company, William Kennedy-Laurie Dickson, 1901) *
Panorama of Crystal Palace Lake (George A. Smith, 1900)
Panorama of London (Hepworth Manufacturing Company, Gaston Quiribet, 1908)
Panorama of London Streets (Embankment) (Paul's Animatograph Works, Robert W. Paul, 1898)
Panorama of Palm Green (Charles Urban Trading Company, H.M. Lomas, 1903)
Panorama of the Tower of London and Tower Bridge (Warwick Trading Company, Will Baker, 1904)
Panorama of Westminster Palace (Warwick Trading Company, Cecil Hepworth, 1898)
Panorama View of Crystal Palace Terraces (George A. Smith, 1900)
Parliament Street (Paul's Animatograph Works, Robert W. Paul, 1896)
A Peep at the Zoo (Walturdaw, 1905)
A Peep round the Zoo (Walturdaw, 1910)
Pelicans (Lumière Brothers, Alexandre Promio, 1896) *

8 "Turnham Green", "Kew Bridge", "Apple Orchards", "Twickenham", "New Power House", "Bushey Park", "Hamton Court", and "Guests Looking Around Marquee".

Petticoat Lane (1903) *
Petticoat Lane (Robert W. Paul, 1896)
Piccadilly Circus (Lumière Brothers, Alexandre Promio, 1896) *
Pleasure Boats on Lake of St James's Park (Warwick Trading Company, Cecil Hepworth, 1898)
The Prehistoric Monsters at Crystal Palace (George A. Smith, 1900)
President Loubet's Visit to London (Warwick Trading Company, Will Baker, 1903)
Queen Victoria's Diamond Jubilee (Paul's Animatograph Works, Robert W. Paul, 1897)[9] *
Railway Traffic on the London and North Western Railway (Charles Goodwin Norton, 1897)
A Ramble Around the Zoo (Charles Urban Trading Company, 1900)
A Ramble through the Zoo (B&C, 1913)
Ride Through Piccadilly Circus (Warwick Trading Company, 1901)
Rotten Row, Hyde Park (Warwick Trading Company, Cecil Hepworth, 1898)
Round London with a Camera (Charles Urban Trading Company, 1910)
Rush Hour Record on London Bridge (Robert W. Paul, 1896)
Scene at the London Docks (British Gaumont, 1905) *
Scenes at the London Zoo (Warwick Trading Company, Cecil Hepworth, 1898) *
Scenes Taken from a Moving Train (British Cinèmatograph, 1897)
Shooting the Chutes (Charles Urban Trading Company, H.M. Lomas, 1903)
Sirdar's Reception at Guildhall (Paul's Animatograph Works, Robert W. Paul, 1900)
Snapshots of the Zoo (Kineto, 1912)
South African Products Exhibition (Walturdaw, 1907)
Spitalfields Market (British Gaumont, 1906)
St Martin in the Fields (Warwick Trading Company, 1897)
Street Hawkers (Paul's Animatograph Works, Robert W. Paul, 1904)
The Streets of London (Charles Urban Trading Company, 1906)[10]
The Streets of London (Kineto, 1912)
The Suffragette Demonstration (Warwick Trading Company, 1908) *

9 "Head of the Procession Including Bluejackets", "Head of the Colonial Procession, Canadians etc.", "Royal Carriages Passing Westminster", "Cape Mounted Riflemen Passing St Paul's", "Dragoons Passing St Paul's", "Royal Carriage Arriving at St Paul's", "Royal Princes in St Paul's Churchyard", "Queen's Carriage and Indian Escort Arriving at St Paul's", and "Life Guards and Princes North of St Paul's."

10 "Hyde Park to Tower Bridge", "Tower of London and Whitechapel", "Five minutes in busiest London", "From East London Slums to Piccadilly."

Suffragettes in Hyde Park (Warwick Trading Company, 1908) *
Suffragettes in Hyde Park (Gaumont, 1911)
Sunday in Petticoat Lane (Hepworth Manufacturing Company, Cecil Hepworth, 1905)
Sunday Morning in Petticoat Lane (British Gaumont, 1906)
Swimming and High Diving at East India Docks (Charles Urban Trading Company, 1905)
Swings at the Crystal Palace (George A. Smith, 1900)
The Thames in Winter (Hepworth Manufacturing Company, Gaston Quiribet, 1909)
The Thames: Westminster to Greenwich (Charles Urban Trading Company, 1910)
Tigers (Lumière Brothers, Alexandre Promio, 1896) *
Togo's Heroes at the Crystal Palace (George A. Smith, Charles Urban Trading Company, 1906)
Tower Bridge Working (Northern Photographic Works, Birt Acres, 1896)
Tower Bridge (Lumière Brothers, Alexandre Promio, 1896) *
Trafalgar Day (W. Watson & Sons, 1897)
Trafalgar Day in London (Warwick Trading Company, Cecil Hepworth, 1897)
Trafalgar Square (Wordsworth Donisthorpe and William Carr Crofts, 1890)
Traffic in King's Road, Chelsea (William Friese-Greene, 1890)
The Traffic in London (J.B. McDowell, 1913)
Traffic on Tower Bridge (Robert W. Paul, 1896) *
Tramway Ride from Shepherd's Bush to Southall (Warwick Trading Company, 1903)
Unloading a Cargo at London Docks (Robert W. Paul, 1898)
Visit of President Fallières (Charles Urban Trading Company, 1908)
A Visit to the London Zoo (Wardour, 1914)
A Visit to the Zoological Gardens (Gaumont, 1910)
A Visit to the Zoological Gardens: the Ruminants (Gaumont, 1911)
A Visit to Westminster Abbey (B&C, J. McDowell, 1911)
The West End of London (Hepworth Manufacturing Company, Gaston Quiribet, 1909)
Westminster Bridge (Lumière Brothers, Alexandre Promio, 1896) *
Westminster Bridge (Robert W. Paul, 1896) *
Westminster Bridge (Esmé Collings, 1896)
Westminster Steam-Boat Pier (Warwick Trading Company, Cecil Hepworth, 1898)
Who's Who at the Zoo (Gaumont, 1910)
The Zoo Series (Walturdaw, 1905)

Liverpool

Church Street (Lumière Brothers, Alexandre Promio 1897) *
Cunard Mail Steamer Lucania Leaving for America (Mitchell and Kenyon, 1901) *
Entrèe dans Clarence Dock (Lumière Brothers, Alexandre Promio 1897) *
The Electric Trains of the Liverpool Overhead Railway (Warwick Trading Company, 1899)
Employees Leaving Alexandra Dock (Mitchell and Kenyon, 1901) *
Everton v. Newcastle United (Mitchell and Kenyon, 1901) *
Everton v. Liverpool (Mitchell and Kenyon, 1901) *
Ferries Arriving at Landing Stage (Paul's Animatograph Works, Robert W. Paul, 1897)
Glimpses on the Liverpool Overhead Railway (Warwick Trading Company, 1899)
La rade (Lumière Brothers, Alexandre Promio 1897) *
Lime Street (Lumière Brothers, Alexandre Promio 1897) *
Liverpool as seen from the Front of an Electric Car from the Pierhead to the Circus (Mitchell and Kenyon, 1901) *
Liverpool Church Parade and Inspection (Mitchell and Kenyon, 1901) *
Liverpool from River Mersey (Warwick Trading Company, 1899)
The Liverpool Pageant (Gaumont, 1907)
Liverpool Police Sports (Archer & Sons, 1898)
Liverpool Street Scenes (Mitchell and Kenyon, 1901) *
Liverpool: the Gate of the West (Walter Tyler, 1910)
Liverpool v. Small Heath (Mitchell and Kenyon, 1901) *
Lord Roberts Presenting Medals to Boer War Volunteers in Liverpool (Mitchell and Kenyon, 1901) *
Panorama Pris du Chemin de Fer Electrique (Lumière Brothers, Alexandre Promio 1897) *
The Port of Liverpool (Rosie Films, Joe Rosenthal, 1910)
Prince of Siam aboard the New Ferry (Mitchell and Kenyon, 1901) *
Princess Louise at Liverpool (Mitchell and Kenyon, 1901) *
Return of Liverpool Yeomanry from the Front (Warwick Trading Company, 1901)
Review of the Liverpool Police (Warwick Trading Company, 1901)
RMS Lucania (Mitchell and Kenyon, 1901)
R.M.S. Oceanic (Warwick Trading Company, 1899)[11]
A Trip to North Wales on the St Elvies (Mitchell and Kenyon, 1901) *
SS Saxonia in Liverpool (Mitchell and Kenyon, 1901) *

11 This series included: "Arrival of the Oceanic at Liverpool" and "Panorama of Liverpool."

SS Skirmisher in Liverpool (Mitchell and Kenyon, 1901) *
St George's Day Procession in Liverpool (Mitchell and Kenyon, 1901) *
Trafalgar Day in Liverpool (Mitchell and Kenyon, 1901) *
Visit of Earl Roberts and Viscount Kitchener to Receive the Freedom of the City (Mitchell and Kenyon, 1901) *
Woodside Ferry (Chard's Vitagraph, 1897)
Woodside Ferry (Vitagraph, 1898)

Manchester

Congregation Leaving Manchester Cathedral (Mitchell and Kenyon, 1901) *
Co-operative Wholesale Clothing Factory in Manchester (Mitchell and Kenyon, 1901) *
Employees of Co-operative Wholesale Society Printing Works, Lonsight, Manchester (Mitchell and Kenyon, 1901) *
Employees Leaving Crossley's works, Openshaw, Manchester (Mitchell and Kenyon, 1901) *
Emplyees of Galloway Ltd Boiler Works, Hyde Road, Manchester (Mitchell and Kenyon, 1901) *
Employees of J. & E. Waters, Talbot Mills, Ellesmere St, Manchester (Mitchell and Kenyon, 1901) *
Funeral in Manchester (Mitchell and Kenyon, 1901) *
General Buller's visit to Manchester (Mitchell and Kenyon, 1901) *
Lord Roberts's Visit to Manchester (Mitchell and Kenyon, 1901) *
Manchester (Warwick Trading Company, 1898)
Manchester and Salford Harriers' Cyclists' Procession (Mitchell and Kenyon, 1901) *
Manchester Band of Hope Procession (Mitchell and Kenyon, 1901) *
Manchester Catholic Orphanage Boys (Mitchell and Kenyon, 1901) *
Manchester Catholics' Whitsuntide Procession (Mitchell and Kenyon, 1901) *
The Manchester Cup (Warwick Trading Company, 1898)
Manchester Italian Catholic Procession (Mitchell and Kenyon, 1901) *
Manchester Spiritualists' Procession (Mitchell and Kenyon, 1901) *
Manchester Street Scenes (Mitchell and Kenyon, 1901) *
Manchester Sunday Schools' Whit Monday Procession (Mitchell and Kenyon, 1901) *
Manchester to Blackpool Road Races (Mitchell and Kenyon, 1903) *
Manchester's Welcome to the Imperial Yeomanry (Mitchell and Kenyon, 1901) *
Manchester Wheelers' Annual Race Meet (Mitchell and Kenyon, 1901) *
The Police Annual Inspection at Birchfields (Mitchell and Kenyon, 1901) *

Race for the Muriatti Cup at the Manchester Wheelers' Annual Race Meet (Mitchell and Kenyon, 1901) *
Return of the Brave Manchester Volunteer (Mitchell and Kenyon, 1901) *
Royal Visit to Manchester, Owens College (Mitchell and Kenyon, 1901) *
Scenes in a Manchester Park (Mitchell and Kenyon, 1901) *
Torpedo Boats Passing Through the Manchester Ship Canal (Mitchell and Kenyon, 1901) *
Torpedo Flotilla Visit to Manchester (Warwick Trading Company, 1901)
Workers at Berry's Blacking Works, Manchester (Mitchell and Kenyon, 1901) *
Workers at Brooks and Doxey, West Gorton Works, Manchester (Mitchell and Kenyon, 1901) *
Workmen Leaving Brooks and Doxey, West Gordon Works, Manchester (Mitchell and Kenyon, 1901) *
Workmen Leaving Peacock's Works at Meal Time, Gorton, Manchester (Mitchell and Kenyon, 1901) *
Workpeople Leaving Ryland's Mill, Gorton, Manchester (Mitchell and Kenyon, 1901) *

Other English towns

Bandstand on Weston-super-Mare Pier (1905)
Birmingham (Robert W. Paul, 1896)
The Blackpool Aviation Week (Walturdaw, 1909)
The Blackpool Aviation Week (Paragon, 1909)
The Blackpool Aviation Week (Kineto, 1909)
Blackpool: Panorama of the Sea Panorama of Blackpool (Hepworth Manufacturing Company, Cecil Hepworth, 1904)
Blackpool: The Parade (Warwick Trading Company, Cecil Hepworth, 1899)
Boys Scrambling for Pennies Under the West Pier (Esmé Collings, 1896) *
Bradford Street Scene (Riley Bros, 1898)
Brighton (Warwick Trading Company, George A. Smith, 1898)[12]
The Brighton Fire: Arrival of the Brigade (George A. Smith, 1899)
Brighton Pier (Paul's Animatograph Works, Robert W. Paul, 1897) *
Brighton Seagoing Electric Car George (George A. Smith, 1897)
Carting Snow, Leicester (Paul's Animatograph Works, Robert W. Paul, 1897)

12 This series included: "Children on the sands", "The ungallant lover", "The esplanade", "Launch of the Skylark" and "The lower esplanade."

City of London Imperial Volunteers Embarking (Biograph Syndicate, 1900)
Conway Castle – Panoramic View of Conway on the L & NW Railway (British Mutoscope and Biograph Company, 1898) *
The City of York (Frank Butcher for Empire, 1911)
CIVs Embarking at Southampton (Warwick Trading Company, 1900) *
The Coventry Pageant (Gaumont, 1907)
Crowds at Brighton (Esmé Collings, 1896)
Diving and Bathing in the Sea Donkey (Paul's Animatograph Works, Robert W. Paul, 1897)
Early Fashions on Brighton Pier (James Williamson, 1900) *
Guildford (J.B. McDowell, B&C, 1913)
Historic Oxford (Cricks & Martin, 1911)
Kitchener's Arrival at Southampton (Warwick Trading Company, Cecil Hepworth, 1900)
Leeds Bridge (Louis Le Prince, 1888) *
The Lord Mayor at Southampton (Paul's Animatograph Works, 1900)
Menai Bridge – The Irish Day Mail from Euston Entering the Tubular Bridge Over the Menai Straits (British Mutoscope and Biograph Company, 1898) *
Nottingham Goose Fair (Warwick Trading Co., 1900)
Panorama of Plymouth Harbour (Warwick Trading Company, Cecil Hepworth, 1898)
Panorama of the City of Bristol (Hepworth Manufacturing Company, 1904)
Portsmouth: the Ferry (George A. Smith, 1898)
Portsmouth Fire Brigade (Chard's Vitagraph, 1898)
The Promenade at Brighton (Esmé Collings, 1896)
Quaint Kentish Towns: Canterbury (Urban Trading Company, 1911)
The Queen at Sheffield (Riley Bros, 1898)
Riding on the Beach (Paul's Animatograph Works, Robert W. Paul, 1897)
Salisbury to Plymouth (Charles Urban Trading Company, 1908)
Soldiers in Portsmouth (Esmé Collings, 1896)
A Street in Brighton (Esmé Collings, 1896)
A Street in Portsmouth, Portsmouth: the Ferry Soldiers in Portsmouth (Esmé Collings, 1896)
Street Scene in Boar Lane (Riley Bros, 1898) *
Street Scene in Brighton (William *Friese-Greene, 1888*)
Street Scenes (Paul's Animatograph Works, Robert W. Paul, 1898)[13]

13 This series included: "Brighton", "Edinburgh", "Sheffield", "Hull", "Liverpool" and "Cardiff."

Tram Journey Through Southampton (British Mutoscope and Biograph Company, 1900) *
Wolverhampton Exhibition (Biograph Syndicate, 1902)
Worcester Street (Philipp Wolf for David Devant, 1898)
The York Pageant (Louis N. Parker, Charles Urban Trading Company, 1909)

Scotland

Glasgow

Edward VII Lays Foundation Stone (1903)
Glasgow (Paul's Animatograph Works, Robert W. Paul, 1896)[14]
Glasgow and the Clyde Coast (Kineto, 1909)
Glasgow Sports (Gaumont, 1903)
Glasgow Trams (Mitchell and Kenyon, 1901) *
Jamaica Street Glasgow (Mitchell and Kenyon, 1901) *
Macmillan's Shop front (1905) *
Response of the Glasgow Tramway Men to the Country's "Call to Arms" (1914)
Photographic Convention (1898) *

Other Scottish towns

Aberdeen (Walker's Royal Cinematograph, 1906) *
Aberdeen Bathing Station Scenes (William Walker, 1898)
Aberdeen Police Sports, Fire Light Hoarse (William Walker, 1898)
Aberdeen South Breakwater in a Storm (William Walker, 1898)
Aberdeen University Quarter Centenary Celebrations (Paul's Animatograph Works, 1906)
Banchory from the Railway (William Walker, 1898)

14 This series included: "Buchanan Street", "Sauchiehall Street", "Rothesay Pier", "Argyle Street").

Barry and Buddon Camps (William Walker, 1898)
Bonnie Scotland (Charles Urban Trading Company 1906)[15]
Bonnie Scotland (B&C, 1911)
Boys' Brigade Inspection at Yorkhill (Mitchell and Kenyon, 1906) *
Camerons at the Charge (William Walker, 1898)
Congregation Leaving St Mary's Church, Dumfries (Mitchell and Kenyon, 1901) *
Coronation Parade in Broxburn (Miller/Regal Cinema, 1910)
Curling and Skating at Glenmuick (William Walker, 1898)
Dumfries Street Scenes (1910, no credits available) *
Duttine Park Sports (William Walker, 1898)
Edinburgh (Paul's Animatograph Works, Robert W. Paul, 1898)
The Edinburgh Fire Department (Warwick Trading Company, 1900)
Employees Leaving Baxter's Jute Works, Dundee (Mitchell & Kenyon, 1901) *
Employees Leaving Gilroy's Jute Works, Dundee (Mitchell & Kenyon, 1901) *
Engines of the Metropolitan Fire Brigade (William Walker, 1898)
The First A.V.A. (William Walker, 1898)
First Volunteer Battalion Gordon Highlanders (William Walker, 1898) *
Funeral of Alexander Buist, Dundee (1901)
Funeral of the Great Lafayette (1911) *
Funeral of Provost Young, Stranraer (1914, no credits available) *
Glimpse of Camperdown Work (Baillie John Noble/Oxford Picture House 1912)
Gordon Highlanders (William Walker, 1898)
Gordon Highlanders Battery Firing (William Walker, 1898)
Hawick Scenes (1909–10, no credits available)
Hospital Saturday Cycle and Trades Procession (William Walker, 1898)
The King's State Visit to Scotland (Gaumont, 1903)[16]

15 This series included: "Edinburgh", "London Scottish Volunteers", "Cave dwellers of Caithness", "Railway panorama through Sutherlandshire", "Across Ross-shire on the urban bioscope special", "Rogie falls salmon fishing", "From Kyle of Lochalsh", "From the mainland to St Kilda", "The isle of Harris and its tweed industry", "From Euston to Crewe", "From Crewe to Carlisle", "Rothesay to Oban", "Fingal's Cave", "Iona and the abbey ruins", "Oban to Loch Ness", "Round about Inverness", "Ayr: the Robbie Burns country", "Aberdeen and Balmoral", "Allen Water and Killicrankie", "The Trossachs and Loch Lomond", "Poverty Island: Lewis."

16 This series included: "Procession from Dalkeith House to Holyrood Palace", "The King at Edinburgh Castle" and "The King laying foundation stone, Glasgow Technical College."

The King's Visit to Edinburgh and Glasgow (Warwick Trading Company, 1903)[17]
Lively Pillow Fight by Children (William Walker, 1898)
Love Burn Street School, Dumfries (Mitchell and Kenyon, 1901) *
Lizars Edinburgh (Mitchell and Kenyon, 1901) *
Lord Roberts Visit to Aberdeen (1913)
MacGregor's Gathering on the Clyde (William Walker, 1898)
March of the Cameron Men (William Walker, 1898)
Opening Cooper Park, Elgin (1903, no credits available)
Opening of Glasgow International Exhibition Panorama of Exhibition (Biograph Syndicate, 1901)
Over the Tay Bridge on an Express Train (Mitchell and Kenyon, 1901) *
Panorama of Banchory, Scotland (Warwick Trading Company, 1900)
Procession in Edinburgh (Mitchell and Kenyon, 1904) *
Public School in Annan (Mitchell and Kenyon, 1901) *
Primrose League Gathering at Glen Tana (William Walker, 1898)
Railway Ride over the Tay Bridge (Peter Feathers, 1897)
Rickarton Camp (William Walker, 1898)
Royal Scots Regiment at Edinburgh Castle (Mitchell and Kenyon, 1901) *
The Royal visit to Scotland (Hepwix, Cecil Hepworth, 1903)[18]
Scenes at Balmoral (1906, no credits available)
Scenes in the Western Highlands of Scotland (Claredon, 1913)
Scholars Leaving St Andre's RC School, Dumfries (Mitchell and Kenyon, 1901) *
Scotland (James Williamson, 1902)[19]
Scotland (Gaumont, 1906)[20]
Scots Grey Entering Perth (William Walker, 1898)
Scottish Highlands in Winter (Charles Urban Trading Company, 1907)
Scottish National Exhibition (Gaumont, 1908)
Scottish Scenery (Gaumont, 1914)
Scottish Sharpshooters Leaving Edinburgh (Warwick Trading Company, 1900)

17 This series included: "King entering the city of Edinburgh" and "King leaving municipal building, Glasgow."
18 This series included: "Arrival of King and Queen at Edinburgh", "March past of Cameron Highlanders", "Scenes at Holyrood Palace" and "Presentation of medals to black watch."
19 This series included: "Circular Panorama of Edinburgh", "Circular Panorama of the Forth Bridge" and "Panorama of Princes Street, Edinburgh."
20 This series included: "A trip to the Hooghly", "Rothesay Bay" and "A day in Burns' Land."

Scottish Shepherds at Work (Hepworth Manufacturing Company, Florence Turner, 1914)
SS Mongolia Leaving the Clyde (Mitchell and Kenyon, 1906) *
Street Procession in Annan (1912, no credits available) *
Street Scenes (Paul's Animatograph Works, Robert W. Paul, 1896)
Strike in Dundee (1911) *
Theatre Fire in Edinburgh (1911) *
Through the Highlands (Kineto, 1913)
Trains at Ballater Station (William Walker, 1898)
Tramway Car Procession (William Walker, 1898)
A Trip Through Scotland (Charles Urban Trading Company, 1908)
Turn Out of the Dundee Fire Brigade (Mitchell and Kenyon, 1901) *
Winter Sport in Scotland (Warwick Trading Company, 1900)
Workforce of Pollock's Caledonian Cabinet Works, Beith (Mitchell and Kenyon, 1901) *
Workforce of Scott & Co., Shipyard, Greenock (Mitchell and Kenyon, 1901) *

Ireland

Dublin

Congregation Leaving Jesuit Church of St Francis Xavier, Dublin (Mitchell and Kenyon, 1902) *
Congregation Leaving St Mary's Pro-Cathedral, Dublin (Mitchell and Kenyon, 1901) *
Crowds Watch City Fire Brigade (Lumière Brothers, Alexandre Promio, 1897) *
Dublin Civic Exhibition Opened by the Viceroy and Lady Aberdeen (British Pathé, 1914)
Dublin Horse Show (Biograph Syndicate, William Dickson, 1900)
Dublin Horse Show (Charles Urban Trading Company, 1906)
Dublin Horse Show (London Cinematograph Company, John Y. Brown, 1909)
Dublin Strikers (Pathe's Animated Gazette, 1913)
Her Majesty's Visit to Ireland (Warwick Trading Company, 1900)[21]
In and About Dublin (Kineto, 1910)

21 This series included: "Return of Earl Cadogan" (Henry Arthur Cadogan, Viscount Chelsea, a Conservative MP), "Royal Procession Passing Through Dublin", "21st Lancers Riding Through Dublin", "Lancers, Guards and Volunteers".

In and Around Dublin (Heron, 1913)
The King's Visit to Ireland (Gaumont, 1904)[22]
Panorama of College Green, Dublin (Mitchell and Kenyon, 1902) *
The Queen's Entry into Dublin (Northern Photographic Works, Birt Acres, 1900)
The Queen's Visit to Dublin (Hepwix, Cecil Hepworth, 1900)[23]
Queen Victoria's Visit to Ireland (Biograph Syndicate, William Dickson, 1900)[24]
Royal Visit to Dublin (Hepworth Manufacturing Company, Cecil Hepworth, 1904)[25]
Scenes of Irish Life at Dublin (Vitagraph, 1913)

Belfast

Belfast Street Scenes (1898) *
Queen's Bridge (Lumière Brothers, Alexandre Promio, 1897) *
Review of Ulster Division at Malone Park by Sir H. Calmont (1914)
Ride on a Tramcar Through Belfast (Mitchell and Kenyon, 1901) *
Scenes of Belfast Riots (Gaumont, 1907) *
Sir Edward Carson and Lord Londonderry Outside the Committee Rooms in Belfast (Gaumont, 1912)

Cork

Albert Quay in Cork (Mitchell and Kenyon, 1902) *
Arrival of VIPs for Official Opening of Cork Exhibition (Mitchell and Kenyon, 1902) *
Cork and Vicinity (Vitagraph, 1913)
Congregation Leaving St Mary's Dominican Church in Cork (Mitchell and Kenyon, 1902) *

22 This series included: "Arrival and Procession, Dublin", "King and Queen at Belfast", and "Arrival of King and Queen at Kingstown and Dublin."
23 This series included: "Late Queen's visit to Dublin: Royal Procession Entering the City Gates."
24 This series included: "Lord Lieutenant of Ireland Entering Phoenix Park", "The Queen Entering Phoenix Park" and "Presenting Bouquet to Queen in Phoenix Park."
25 This series included: "Arrival at Kingstown and Dublin", "Levee the Dublin", and "Royal Party at Belfast."

Congregation Leaving St Patrick's Church in Cork (Mitchell and Kenyon, 1902) *
Lord Mayor of Cork arriving for Official Opening of Cork Exhibition (Mitchell and Kenyon, 1902) *
Panorama of Cork Exhibition Grounds (Mitchell and Kenyon, 1902) *
Preparation of the Cork Exhibition Grounds and Erection of Buildings (Mitchell and Kenyon, 1902) *
Prince Henry of Prussia to Cork Exhibition (Mitchell and Kenyon, 1902) *
Trade Procession at Opening of Cork Exhibition (Mitchell and Kenyon, 1902) *
Tram Ride from King Street to Patrick's Bridge, Cork (Mitchell and Kenyon, 1902) *
Views of the Grand Parade, Cork (Mitchell and Kenyon, 1902) *
The Visit of the Duke of Connaught C-I-C Forces in Ireland (Mitchell and Kenyon, 1902) *

Other Irish towns

The Bioscope in Ireland (Warwick Trading Company, 1900)
A Cattle Drive in County Galway (Paul's Animatograph Works, 1908)
Cattle Market in Derry (Mitchell and Kenyon, 1902) *
Departure of a train at Amiens Street Station (Lumière Brothers, Alexandre Promio, 1897) *
Derry Election Day (Topical Film Company, 1913)
Hussars Passing Through Dublin (Lumière Brothers, Alexandre Promio, 1897) *
In Ireland: Excursion to Killarney (British Pathé, 1908)
Irish Life and Character (Warwick Trading Company, 1904)
Irish Life: Pig Raising (General, 1913)
Irish Scenes and Types (Charles Urban Trading Company, 1906)
An Irish Stag Hunt (Warwick Trading Company, 1902)
Life in Wexford (Mitchell and Kenyon, 1902) *
Regiments Returned from the Boer War to Victoria Barracks, Cork (Mitchell and Kenyon, 1902) *
Sackville Bridge (Lumière Brothers, Alexandre Promio, 1897) *
Typical Scenes in Ireland (Warwick Trading Company, 1904)
Wexford Railway Station (Mitchell and Kenyon, 1902) *
With the Bioscope through Ireland (Warwick Trading Company, 1900)[26]

26 This series included: "Royal Avenue, Belfast", "Panorama of the Exhibition Grounds, Cork", "Patrick Street Cork", and "High Street, Belfast."

Empire

South Asia

Calcutta (John Montagu Benett-Stanford, 1899)[27] *
Delhi (British Pathé, 1909) *
The Delhi Durbar (Warwick Trading Company, 1903) *
The Delhi Durbar (Gaumont, 1911)
The Delhi Durbar (Charles Urban Trading Company, Charles Urban, 1911) *
The Delhi Durbar (Pathé, 1911) *
The Delhi Durbar (Warwick Trading Company, 1911) *
The Delhi Durbar (Baker Motion Photography, 1911)
The Durbar at Delhi (Paul's Animatograph Works, 1903) *
Historic Mutiny Sites (1914) *
India (Charles Urban Trading Company, 1908)[28]
Life in Hyderabad (Warwick Trading Company, J. Gregory Mantle, 1903)[29]
Missionary Life in India (Walturdaw Company, 1907)[30]
Native Life in Ceylon (Hepworth Manufacturing Company, 1909)*
Our Eastern Empire (Warwick Trading Company, F.B. Steward, 1900)[31]
Scene in the Bazaar at Peshawar (Paul's Animatograph works, 1903)

East Asia

Allied troops taking Chinese Prisoners over Tientsin Bridge (Walter Gibbons, 1900) *
American Transport Entering Peking (Warwick Trading Company, Joe Rosenthal, 1900)
Boxers Decapitating a Prisoner (Walter Gibbons, 1900) *

27 Including "Arrival of a train at Haora station" and "Panorama of Calcutta."
28 This series included: "Mohammedan Festival at Calcutta", "Scenes in Benares, Delhi, Agra", "Madras", and "Through the native quarter, Bombay."
29 This series included: "Panorama of street showing Hindu boy preaching" and "Street scene in Hyderabad."
30 Including "Native Street in India."
31 This series included: "Mid-day siesta in Calcutta", "Church Gate Street in Bombay", "Street Scene in Poona" and "Native street scene in Calcutta."

The Burning of a Missionary (Walter Gibbons, 1900) *
China (Warwick Trading Company, Joe Rosenthal, 1900)[32]
Circular Panorama of Singapore and Landing Stage (Warwick Trading Company, Joe Rosenthal, 1900)
Entry into Peking of Count Von Waldersee (Warwick Trading Company, Joe Rosenthal, 1900)
Outside the walls of Peking (Walter Gibbons, 1900)
Nanking Road, Shanghai (Warwick Trading Company, Joe Rosenthal, 1900) *
In Quaint Peking (Charles Urban Travelling Company, 1911) *
Panorama of Hong Kong Harbour (Ernest Hatch, 1900)
Panorama of Singapore Sea Front (Warwick Trading Company, Joe Rosenthal, 1900)
Street in Tientsin after Occupation by Allied Troops (Warwick Trading Company, Joe Rosenthal, 1900)
Streets and Ruins of Tientsin (Warwick Trading Company, Joe Rosenthal, 1900)
Streets of Tientsin after the Town's Capture (Warwick Trading Company, Joe Rosenthal, 1900)
Street Scene in Peking (Ernest Hatch, 1900)
The Urban-Bioscope Expedition into Borneo (Charles Urban Trading Company, Harold Mease Lomas, 1904)
Volunteer Turnout at Shanghai (Ernest Hatch, 1900)

32 This series included: "Busy Street in Shanghai", "Sikhs' Camp in Shanghai", "Chinese junks in Hong Kong Harbour", "Circular Panorama Hong Kong Harbour", "Chinatown Bazaar – Hong Kong", "Ride on the Peak Tramway", "Lively Street Scene in Shanghai", "Curious Natives on Shanghai Streets", "Circular Panorama of Native Quarters", "Foreign Warships off the Bund", "Panorama of the Bund", "Chinese Market and canal", "Street in Tientsin after Occupation by Allied Troops", "Streets and Ruins of Tientsin", "Native Quarters of Tientsin", "Panorama of Tientsin from the River", "Street Life in the Tartar City", "Panorama of Peking from the Tartar Gate", "Camel Transport Entering Peking", "American Transport Entering Peking", "Entry into Peking of Count Von Waldersee", "Jinricksha Stand – Shanghai", "Sea Front Drive of Shanghai", "Nankin Road, Life in a Canton Street", "Queen's Road – Hong Kong", "Shanghai from a Sailing Junk", "Shanghai Shops and Opium Dens", "Review of the Fourth Goorkhas at Hong Kong" and "Streets of Tientsin after the Town's Capture."

Africa

Adderley Street, Cape Town (Warwick Trading Company, Edgar Hyman, 1900) *
Alarming Queen's Company of Grenadier Guards at Omdurman (John Benett-Stanford, 1898)
Arrival and Reception of Lord Roberts at Cape Town (Warwick Trading Company, Edgar Hyman and Joe Rosenthal, 1900) *
The Bioscope in Algiers (Warwick Trading Company, C. Rider Noble, 1902)[33]
The Bioscope in Egypt: Cairo to Khartoum (Warwick Trading Company, C. Rider Noble and Joe Rosenthal, 1901)
Cape Town (Warwick Trading Company, Edgar Hyman, 1900)[34]
Cape Town: Ride Through Adderley Street (Warwick Trading Company, Edgar Hyman, 1900)
Cape Town to Seapoint by Tram (Warwick Trading Company, Edgar Hyman, 1900)
Circular Panorama of Port Said (Warwick Trading Company, C. Rider Noble and Joe Rosenthal, 1901)
CIVs at Cape Town (Gibbons' Bio Tableaux, C. Rider Noble, 1900)
CIVs March on Johannesburg (Warwick Trading Company, 1899)
City Imperial Volunteers Leaving Wellington Barracks (Paul's Animatograph Work, Sydney Melsom, 1900)
Day in the Life of a Rickshaw Boy (Empire, Frank Butcher and Emile Lauste, 1911) *
Egypt (Paul's Animatograph Work, Henry Short, 1897)[35]
Egypt and Her Defenders (Kineto, 1914) *
From Cairo to the Pyramids (Warwick Trading Company, C. Rider Noble and Joe Rosenthal, 1901)
Hoisting the Union Jack at Johannesburg (Warwick Trading Company, 1899)
In and around Tangier (Hepworth Manufacturing Company, H.M. Lomas, 1907)

33 This series includes: "Panorama of Market Square at Biskra", "Panorama of Oldest Mosque in Algeria", "Scenes in Sidi Okba", "Circular Panorama of Market Square", "Scenes in Kairouan", "Snapshots of Algeria", "Panorama from train approaching El Kantara."

34 This series included: "Panorama of Cape Town", "Cape Town docks", "The Sea Front of Cape Town", "Coaling Vessels by Natives", "Panorama of Table Bay", "Panorama of Table Mountain", "Cape Town from Sea", and "Cape Mounted Police."

35 This series included: "Cairo Scene: Selling Water" and "Loading Camels in a Cairo Street."

Jaffa and Jerusalem (Charles Urban Trading Company, F. Ormiston-Smith, 1905)[36] *
Johannesburg (Warwick Trading Company, Edgar Hyman, 1898)
Johannesburg: a Tram Ride in Commissioner Street (Warwick Trading Company, Edgar Hyman, 1898)
Johannesburg Between the Chains (Warwick Trading Company, Edgar Hyman, 1898)
Johannesburg Handicap (Warwick Trading Company, Edgar Hyman, 1898)
Johannesburg: the Golden City (Empire, Frank Butcher and Emile Lauste, 1911)
Kaffirs and Zulus on the Warpath (G.A. Smith, 1899)
Life in Cape Colony (Paul's Animatograph Work, 1898)
Mogador (Hepworth Manufacturing Company, H.M. Lomas, 1907)
Moonlight Trip on the Nile (Hepworth Manufacturing Company, 1909)
Naval Brigade Through Main Street (Warwick Trading Company, 1899)
New South Wales Lancers Landing at Cape Town (Biograph Syndicate, William Dickson, 1899)
Panoramic View of Table Bay (William Dickson, Biograph Syndicate, 1898)
Port Elizabeth (Warwick Trading Company, Edgar Hyman, 1900)[37]
Port Elizabeth Troops and Artillery to the Front (Warwick Trading Company, 1899)
Ride on a Trolley Car in Cape Town (Strand Road) (Warwick Trading Company, Edgar Hyman, 1900)
Savage South Africa at Earl's Court (John Benett-Stanford, 1899)
Savage South Africa: Savage Attack and Repulse (Warwick Trading Company, 1899)
Scenes in and Around Cape Town (Empire, Frank Butcher and Emile Lauste, 1911)
Scenes of Egyptian Life or Land of the Pharaohs (Hepworth Manufacturing Company, 1909)
South African Products Exhibition (Walturdaw, 1899)[38]
The Solar Eclipse (Hepworth Manufacturing Company, 1901)[39]

36 This series included: "Jaffa and its Harbour" and "Jerusalem the Holy City."

37 This series included: "Tram Ride up Russell Road", "Tram Ride Through Main Street", "Climbing Cape Road, Panoramic view of Durban", "Trough Cape Road, Port Elizabeth", and "Down Whites Road, Port Elizabeth."

38 This series included: "Acrobatic Piccaninnies", "Piccaninnies and Lion Cubs", "Taming a Lion", "Parades of the Apes", "Boers Hunting Zebras and Wild Beasts", "The Somersault Ape", "Captain Taylor and His Trained Cinnamon Bear", and "Savages attacking the Square".

39 This series includes: "Street Scenes in Algiers", "Tangier, the Landing Stage", and "Tangier, the Market Place."

Street Scene in Cairo (Warwick Trading Company, C. Rider Noble and Joe Rosenthal, 1901)
Through Cape Town on a Tram (Warwick Trading Company, Edgar Hyman, 1900)
Tram Ride through Cape Town (Warwick Trading Company, Edgar Hyman, 1900)
The Uprising in Morocco (Warwick Trading Company, C. Rider Noble, 1902)[40]
Woodstock to Cape Town by Tram (Warwick Trading Company, Edgar Hyman, 1900)

Canada

Montreal (Northern Photographic Works, Birt Acres, 1895)
Montreal (Warwick Trading Company, Joe Rosenthal, 1901)
Farming in N.W. Canada (Charles Urban Trading Company, 1910)
Panorama of Main Street, Winnipeg (Warwick Trading Company, Joe Rosenthal, 1901)
The Resources of Canada (Charles Urban Trading Company, 1914)
The Wheat fields of Canada (Warwick Trading Company, 1908)

Australia

Bourke Street, Melbourne (Imperium Film in 1911) *
Capital Cities of Australia (Film Sale Agency, 1914)
Sydney (Warwick Trading Company, Joe Rosenthal, 1901)
Sydney Street Scenes (Empire Company, *Emile Lauste*, 1912)

40 This series included: "Street life in the Sacred City of Mequinel" and "Moorish street minstrel performing at Morocco."

Index

Aberdeen 132, 147–150, 152–154, 157, 176, 254, 258, 262, 264, 267, 286–289
Acton (London) 103–104
Adams, Annemarie 51, 263
Adelphi (Saltford) 140
advertising 68–69, 72, 75, 95, 142–143, 157, 178
Africa 38, 52, 63, 76, 83, 120–121, 131–132, 139, 141–142, 151, 153, 157, 161–162, 165, 190, 194, 198, 225, 227–230, 234, 236, 238, 242–243, 252, 254, 257, 259, 261, 267, 294–295
Agra 206, 292
al-Sayyid-Marsot, Afaf Lufti 238, 271
Aldcroft, Derek H. 103, 129, 253, 263
Aldwych 67, 78, 145
Algeria 237, 294
Algiers 237, 294–295
Alhambra Theatre 214
Aliwal North 233
Altick, Richard 210, 218, 249
America 25, 31, 110, 123, 128, 282
Ancoats (Manchester) 140
Anderson, Benedict 54, 245
Andhra Pradesh 206
Animatograph 78, 81, 153, 275–280, 282, 284–287, 289, 291–292, 294–295
Annan 148, 288
Annan, Thomas 5, 156, 249
Appleton, R.J. 76
Archer & Sons 127, 282
Archer, Mildred 203, 249
Armes, Roy 20, 24, 249
Arnott, John 178
Arscott, Caroline 49, 62, 70, 156, 263
Ashby, Justine 21–22, 249
Ashcroft, Bill 200, 249
Ashton, Thomas S. 141, 249
Atkins, Tommy (character) 28
Attridge, Derek 174, 265
Attridge, Steven 25, 27, 249
Australia 139, 157, 201, 296
Axminster 120
Ayr 149, 287

Baedeker 53, 58, 62, 249
Bailey, Peter 28, 249
Baillie, John 150, 287
Ballantyne, Tony 249
Ballater 148, 289
Balmoral 148–149, 153, 287–288
Banchory 148–149, 286, 288
Bank of England 54, 56, 58, 67–68, 75–76, 102, 106, 110, 133, 174, 275, 277–278
Baratay, Eric 90, 249
Barber, Stephen 13–14, 250
Barker, Will 47, 57, 60, 62–63, 81, 94, 103, 107, 129, 214–215, 263, 276, 278–280, 292
Barnes, John 43, 78–80, 105, 230, 250
Barnet (London) 105, 276
Barringer, Tim 96, 191, 219, 250, 263, 268
Barsan, Richard 19, 250
Bath 156
Becker, Francis 34, 269
Beckson, Karl 72, 250
Bedford 178

Beevor, Walter C. 230
Beith 148, 289
Belfast 72, 126, 131, 169, 171–172, 175–179, 182, 254, 290–291
Bell, Fred 123–124, 250
Benares 206, 292
Bender & Co. 100, 276
Benett-Stanford, John M. 206–207, 227, 230, 292, 294–295
Bengal 92, 204, 214
Berman, Marshall 14, 92–93, 250
Bernstein, Matthew 35, 250
Bertozzi, Marco 17, 270
Betjeman, John 153, 250
Bhabha, Homi K. 198, 250
Bickford-Smith, Vivian 235, 250
bicycles 43–45, 58, 87, 93–94, 144, 178–179, 224, 255, 275, 277
Bioscope 174, 178, 180, 222, 237, 240, 291, 293–294
Birchfields 143, 283
Birkenhead 125–126, 128
Birmingham 94, 116–117, 284
Birt Acres 59, 63, 76–77, 100–101, 105, 175, 200, 275–276, 281, 290, 296
Bishopsgate 78, 230, 250
Biskra 237, 294
Black, Barbara 95, 250
Blackburn 22
Blackfriars 63, 100, 275
Blackpool 1, 118–119, 143, 283–284
Blackrock 181
Blake, William 50, 253
Bloody Sunday (1886) 65
Boerograph 231
Bombay 55, 80, 194, 205–206, 255, 258, 292
Bonett, Helena 55
Booth, Charles 5, 48, 140–141, 250
Bootle 122
Boots 178
Bordwell, David 20, 250
Borneo 76, 80, 222, 293
Boston 89, 128, 261
Bottomore, Stephen 22, 213–215, 263, 270
Boudica 11, 252
Bourne, Samuel 194
Bowrington Canal 224
Boxer Rebellion 28, 33, 217, 220–224, 229, 231, 242, 260, 292
Boyer, Christine 15, 251
Brabazon, Reginald 165
Bradford 116–118, 284
Bradley, Simon 68, 251
Brady, Joe 173, 251
Brantlinger, Patrick 32, 251
Breckenridge, Carol 203, 270
Briggs, Åsa 6, 117, 139, 201, 251
Brighton 1, 19–20, 44, 115, 118–119, 221, 254, 284–285
Bristol 117, 285
Britannia 33, 174
British Museum 239
British Mutoscope and Biograph Company (Biograph Syndicate) 9, 11, 65, 68, 72, 76, 91, 96, 100, 104, 106, 117, 120, 132, 159, 169, 174–175, 230, 232, 234, 251, 254, 276–279, 285–286, 288–290, 295
Brooks 135, 284
Brown, John Y. 94, 174, 279, 289
Brown, Richard 104, 251
Brown, Scott 240
Brown, Simon 270
Broxburn 148, 287
Bruno, Giuliana 9, 15, 17, 97, 102, 189, 251, 270
Brunsdon, Charlotte 17, 251
Buckingham Palace 54, 76, 82, 140, 252
Buckner, Phillip 200, 251, 270
Buist, Alexander 150, 287

Buller, Redvers 140, 142, 229, 283
Bundi 194
Burch, Noël 4, 21, 24, 26, 221, 251
Burford's Panorama (Leicester Square) 210, 218, 239
Burke, Thomas 175
Burma 194
Burrows, Jon 17, 27, 271
Burton, Antoinette 85, 271
Bushey 101, 104, 279
Butcher, Frank 117, 236, 285, 294–295

Cadogan, Arthur 175, 289
Cairns, Hugh 168
Cairo 239–242, 294, 296
Caithness 149, 287
Calcutta 80, 194, 205–208, 214–215, 251, 255, 292
Calverley Beevor, Walter 230
Cambodia 219
Camden vii 55, 59, 105–106, 278
Canada 75–76, 82–83, 121, 139, 157, 179, 200, 251, 270, 296
canals 122, 134, 139–140, 155, 163, 223–224, 238, 284, 293
Canberra 32, 56, 262, 269
Cannadine, David 11, 30–31, 47, 211, 251, 264
Canosa, Michele 264
Canterbury 117–118, 285
Canton (Guangzhou) 218, 223, 293
Cape Town 59, 79, 132, 228, 232–236, 242–243, 250, 255, 263, 280, 294–296
Cardiff 119
Carlisle 149, 287
Carroll, Clare 166, 251, 269
cars 94–95, 102, 145, 236
carts 13, 43, 52, 87, 127, 138, 160–161, 179, 204, 217
Cavendish, Frederick 175
Çelik, Zeynep 51, 251, 263
Ceylon 205–207, 255, 292
Chanan, Michael 20–21, 54, 80, 188, 231, 251
Chaplin, Charles 27
Charing Cross 56, 64, 66, 77, 81, 100, 102, 277–278
Charles Urban Trading Company 242, 275–281, 285–289, 291–293, 296
Charlottesville 5, 95, 250, 262
Chattopadhyay, Swati 207, 251
Chelsea 64, 108, 175, 281, 289
Cheng, Vincent J. 174, 251
China 20, 28, 33, 139, 194, 216–224, 229, 234, 242, 260, 268, 293
Chiswick (London) 103
Chowdhry, Prem 202, 251
Christie, Ian 44, 56, 251
Cinquegrani, Maurizio 55
City Imperial Volunteers (CIV) 53, 58, 74, 131–132, 232, 294
Clapham, Adam 246
Claredon 149, 288
Clark, Peter 49, 68, 123, 156, 252, 259, 263–264, 268
Clarke, David 15, 252
Clercq, Louis 173
Clifford, James 196, 252
Cobh 128
Cochran, Sherman 224, 252
Cohn, Bernard S. 209, 264
Collingridge, Vanessa 11, 252
Collings, Esmé 63, 70, 118, 120, 276, 281, 284–285
Colombo 187, 205, 209, 216
Conder, Josiah 207, 252
Condon, Denis 168, 264
Connor, John 107
Constantinople 189
Conway 100, 285
Cook, Thomas 192

Coombes, Annie 227, 252
Coppola, Francis Ford 50
Cork 72, 165–166, 171–173, 179–182, 251–252, 257, 269, 290–291
Corkin, Stanley 109–110, 252
coronations 1, 61, 72, 74, 81–83, 94, 148, 211–212, 214–215, 265, 276, 287
Cosgrove, Denis 36, 57, 86, 187, 252
Costa, Antonio 7, 264
Coulthurst, Samuel 5, 135, 156, 258
countryside 37, 74, 118, 147–149, 166, 171, 176–177, 205
Couteur, John 77
Coventry 102, 117, 285
Cowes 81
Cowman, Krista 124–125, 129, 264
Crewe 149, 287
Cricks & Martin 117–118, 285
Crimean War 58, 63, 69, 231
Crofts, William C. 64, 281
Crook, William 204, 271
Crossick, Geoffrey 25, 252, 268
crowds 52, 58, 69–70, 72, 74–75, 119–120, 132–133, 143, 152, 169, 210, 241, 267, 285, 289
Crowley, John 181, 252
Croydon (London) 100, 126, 276
Cruickshank, George 48
Crystal Palace 1, 54, 56, 59, 85, 89, 92–95, 110, 152, 187, 208, 246, 266, 275–277, 279–281
Curzon, George N. 213

Daiches, David 156, 160, 252
Daly, Nicholas 231–232, 252
Daniells 198
Daniels, Stephen 51, 252
Dartmouth 120
Daunton, Martin 49, 68, 252, 259, 263–264, 266
Davies, Terence 246
Davis, Andrew 140, 252
Davis, Mike 14, 220, 252
Davison's Film Agency 201
Deansgate 137–138, 140, 145
Dehejia, Vidya 194, 252
Delhi 22, 29, 187, 200–201, 206, 208–216, 242–243, 253–254, 263, 265, 270, 272, 292
Delhi Durbars 22, 200, 208–209, 211–216, 242, 265, 270, 292
demonstrations 6, 25, 58–59, 64–66
Denny, Alice 238, 261
Derby 122, 127–128
Derry (Londonderry) 171, 177, 179, 182, 290–291
Devon 118
Dickens, Charles 1, 5, 48, 172
Dicks, Brian 154, 264
Dickson, William 23, 65, 68, 72, 76, 81, 91, 95, 104–105, 107, 159, 174–175, 230, 232, 234, 276–279, 289–290, 295
Dixon, Bryony 22–23, 103, 252, 260, 263, 266
Dixon, Wheeler W. 22, 252, 263
Dobson, Denis 87, 257
docks 46, 98, 121–128, 130, 134, 150, 158, 176, 179–180, 219, 232, 234–235, 237, 254, 268, 273, 280–282, 290, 294
Donald, James 16, 253
Donisthorpe, Wordsworth 64, 281
Doré, Gustave 48
Dover 207
Dowey, W. & D. 77
Doxey 135, 284
Doyle, Arthur Conan 1
Driver, Felix 36, 46–47, 74–75, 87–88, 96, 98, 192–193, 253, 264, 266–268, 271
Dublin 55, 72, 165, 168–169, 171–179, 182, 243, 251, 255, 264–265, 267, 272, 289–291

Dudley, Robert 121, 124
Dumfries 148, 287–288
Dundee 147, 150, 287, 289
Durban 234, 236, 295
Durham 193, 254

Ealing (London) 23, 103–106, 109, 163, 279
Edinburgh vii 48, 119, 147, 149–153, 250, 253, 258, 261, 285, 287–289
Edison, Thomas 8, 43, 110, 170
Edwards, Brian 150, 168, 253
Egypt 88, 182, 189, 201, 236–243, 256, 259, 265, 271, 294
Eight-Nation Alliance 220, 223–224
Elgin 148, 288
Elsaesser, Thomas 9, 266
Embankment 54, 56, 62, 239, 279
Engels, Friedrich 52, 136, 145, 253
England 1, 16, 20, 32, 43, 51–52, 54–56, 58, 67–68, 73, 75–76, 78, 80, 95, 105, 110, 115–118, 131, 133, 135–136, 143, 145, 170, 173, 189–190, 210, 218, 221, 227, 230, 240, 245, 250, 252–253, 256, 259–262, 269, 271, 273, 275, 277–278
Erdman, David 50, 253
Ettrick Bay 159
Evans, Mortimer 115
Everton 122, 126, 282
exhibitions 1–2, 19–20, 27, 31, 34, 37–38, 44, 46, 51, 59, 82, 85–89, 92, 94–97, 108, 117, 151–152, 158–159, 162, 169, 171, 176, 179–182, 187–188, 190, 193, 203–204, 212, 216, 227–228, 231, 238–239, 243–244, 246, 254, 257, 263, 267, 271, 279–280, 286, 288–291, 295
exoticism vi 38, 166, 188, 191, 197, 200–202, 204, 207–208, 213, 215–217, 219, 221–223, 225, 228, 236, 241, 243, 261, 267–268

Fabianism 30, 262
factories 44, 59, 124, 135–136, 142–143, 150, 251, 266, 278, 283
fairs 9, 27, 31, 85–86, 88, 105, 117, 152, 170–171, 182, 187–188, 191, 203, 205, 254, 270–271, 276, 285
Fallières, Armand 81, 281
famine 131, 166, 179, 220, 234
farming 171, 177, 200, 296
Favro, Diane 51, 251, 263
Feldman, David 71, 265
Fell, John 233, 253, 265
ferries 61–62, 120, 125–126, 128, 282–283, 285
festivals 56, 130, 134, 204, 206, 208, 246, 292
Fez 194
Fife 148, 159
Fifeshire 154
Fillis, Frank E. 228
Film Sale Agency 201, 296
Fingal 149, 287
Finsbury (London) 100, 106, 276
fishing 149, 180, 205, 237, 287
Fitzgerald, Edward 181
Fitzjames Stephen, James 135
Fitzmaurice, Tony 14, 39, 262, 269
Fitzpatrick, David 167–168, 265
Flynn, Tom 96, 219, 250, 263, 268
Foley, John Henry 63
Fordham, Douglas 191, 250
Forestier, Amédée 80
Fornaioli, Enrico 264
Foster, Robert F. 166, 173, 178, 180, 253
Foucault, Michel 14, 34, 190, 253
Fraser, James 198
Fraser, W. Hamish 33, 156, 253, 265, 268–269
Freeman, Michael J. 103, 129, 253, 263
Friese-Greene, William 43, 108, 115–116, 277, 281, 285
Frith, Francis 239

Frykenberg, Robert E. 214, 253, 265
Fullerton, John 17, 253, 265
funerals 72–73, 80–81, 141, 148, 150, 152, 176, 222, 276, 283, 287

Gad, Urban 17
Galaup, Jean François de 193
Galbraith, John S. 238, 271
Gallagher, John 238, 261
Galway 171, 182, 291
Ganges 207
Garibaldi, Giuseppe 58
Gaumont 57, 59, 72, 80, 91, 100, 117, 130, 149, 151, 155, 175–176, 178–179, 213–215, 276–278, 280–282, 285–288, 290, 292
Gaylor, David 148
Gerard, Kearns 256
Gibb, Andrew 156, 253
Gibbons, Luke 168, 261
Gibbons, Walter 221, 230, 232, 292–294
Gibbons' Bio Tableaux 232, 294
Gibson, Edward 168
Gifford, Denis 4, 19, 43, 64, 81, 108, 115, 253–254, 275
Gilbert, David 36, 46–47, 71, 74–75, 77, 96, 98, 110, 253–254, 264–267, 269, 271
Gilfillan, John B.S. 161, 258
Gillespie, James 265
Ginner, Charles 55
Gissing, George 5
Giza 241
Gladstone, William E. 80–81, 276
Glasgow 1, 37, 39, 116, 123, 131, 145, 147, 149–151, 154–162, 187, 234, 249, 252–253, 257–259, 264, 267–269, 273, 286–288
Glenmuick 148, 287
Goldman, Sydney 230
Goldstein, Emmanuel 1, 3
Gorky 43–44, 273
Gramsci, Antonio 34
Gray, Frank 118, 254
Gray, John 176, 254
Green, George 170
Greenhalgh, Paul 86, 254
Greenock 148, 289
Greenwich 62, 281
Grewal, Inderpal 193, 254
Grierson, John 245
Griffiths, Alison 191, 254
Grottle Strebel, Elizabeth 233, 265
Guangzhou *see* Canton
Guildford 117–118, 285
Guildhall 53, 74, 81, 232, 280
Gunn, Simon 125, 137–138, 140–142, 254, 264–265
Gunning, Tom 8–11, 88, 135, 254, 266, 271
Gupta, Narayani 211–212, 254
Gutsche, Thelma 229, 254

Hackforth-Jones, Jocelyn 199, 254
Haghe, Louis 239
Haldane, Bert 18–19, 107
Hammersmith 103
Hampstead 105, 276
Hampton Court 104
Hansen, Peter 266
Hardouin-Fugier, Elisabeth 90, 249
Hardy, Florence E. 254
Harper, Sue 245, 254
Harrison Company 222
Hartley, Jesse 121
Harvey, David 14, 254
Hassam, Andrew 92, 266
Hatch, Ernest 222, 255, 293
Hawick 148, 287
Hays, Michael 35, 271
Headingly, Adolphe Smith 49, 100, 262
Henderson, Fiona 110, 265

Hepwix 117, 276–277, 288, 290
Hepworth Manufacturing Company 102, 205, 237, 240, 276, 279, 281, 284–285, 289–290, 292, 294–295
Hepworth, Cecil 19–20, 24, 26, 55, 59, 61–62, 65, 68, 72, 76, 81–83, 91–92, 95, 100, 102, 117–120, 149, 151, 169–170, 175, 178–179, 205, 233, 237, 240, 255, 276–277, 279–281, 284–285, 288–290, 292, 294–295
Herbert, Reginald 255
Herlihy, David 44, 255
Heron 174, 290
Hicks, George E. 13, 69
Higson, Andrew 21–22, 231, 249, 255, 268
Hill, John 168, 171, 255, 261, 264, 267
Hitchcock, Alfred 245, 256
Hobhouse, Emily 233–234, 255
Hobsbawm, Eric 3–4, 11, 18, 26, 31, 36, 38, 209, 255, 264
Hobson, John A. 27, 233, 255
Hodder, C. 71, 255
Hodges, William 198
Hoffenberg, Peter 86, 255
Holborn 67
Holt, Ysanne vii 55
Holyhead 100
Holyrood Palace 151, 287–288
Hooghly 149, 288
Horner, Thomas 57
Horrall, Andrew 81, 94
Houses of Parliament 54, 58, 61–62, 76–77, 82–83, 90, 110, 153, 167–168, 174, 277, 279
Hove 118, 221, 254
Howe, Stephen 166, 168, 255
Howes, Marjorie 174, 249, 265
Hubsch 57, 276
Hudson Soap 178
Hughes, W.C. 77
Hull 119, 285
Humboldt, Alexander 193
Hunt, Tristram 116, 130, 171, 256
Hyderabad 206, 292
Hyman, Edgar 230, 232, 234–235, 294–296

Imperial Yeomanry 131, 142, 154, 282–283
Imperium Film 201, 296
India vi 1, 11, 22, 28, 38, 91–92, 121, 123, 131, 139, 157, 159, 163, 167, 187, 189, 191, 193–195, 197–216, 236, 243, 249, 251–252, 255–256, 258, 260, 264, 267, 270–271, 281, 292
industry v 3, 6, 11, 20–21, 25, 31, 38, 43, 59, 74, 85, 95, 111, 115–119, 121–123, 125, 127, 129, 131, 133–137, 139–141, 143, 145, 147, 149–150, 154–156, 158, 161–163, 167, 176, 179, 181–182, 191, 231–232, 252, 254–255, 257, 259, 261, 266, 268, 287
Ingersoll, Richard 51, 251, 263
Inverness 149, 287
Iona 149, 287
Ipswich 83
Ireland 38, 65, 128, 131, 143, 163, 165–171, 173–182, 204, 230, 251, 253, 255–257, 259, 261–262, 264–269, 271, 289–291

Jackson, Alvin 168
Jackson, Gordon 122, 266
Jacob, Steven 245, 256
Jaffa 237, 295
Jamaica 157, 160–161, 286
James Wyld's Globe (Leicester Square) 1, 187
Jameson, Fredric 14, 256
Java 93, 276–278

Jeffery, Keith 165–167, 172, 256, 266, 268
Jenkins, Paul 150, 253
Jennings, Humphrey 245–246
Jenny Deans (locomotive) 101
Jerusalem 237, 242, 295
jingoism 25, 27, 35, 165–166, 233, 255, 268
Johannesburg 189, 232, 234–236, 242, 294–295
Joyce, James 165, 172–174, 182, 256
Joyce, Patrick 66, 109, 116, 256
Judd, Denis 74, 210, 229, 256

Kairouan 237, 294
Kalfatovic, Martin R. 239, 256
Kandy 205
Kantara 237, 294
Kearns, Gerard 256
Keiller, Patrick vii 23, 103–104, 266–267
Kensington (London) 46, 51, 91, 95–96, 208, 217, 243, 263, 277
Kenyon, Sagar 22–24, 118–119, 126–133, 135, 138–144, 148, 150–151, 157–158, 160, 165, 170–173, 177–182, 230–231, 262, 266, 268, 270, 282–284, 286–291
Kern, Stephen 2, 44, 256
Khartoum 240, 294
Kidd, Alan 134, 137–138, 140, 256, 265, 267
Kilkenny 72
Killarney 171, 291
Kilvert, Francis 124, 260
Kinematograph 77–78, 148, 271
Kinesigraph 64
Kineto 57, 91, 119, 149, 159, 169, 174, 241, 280, 284, 286, 289, 294
Kinetoscope 43
Kingstown 175, 290
Kingsway (London) 67, 78, 145
Kipling, Rudyard 28
Kirby, Lynne 99, 101–102, 257
Kitchener, Horatio Herbert 81, 83, 120, 131–132, 229, 240–241, 283, 285
Koeck, Richard vii 125, 271
Koven, Seth 59, 257
Kracauer, Siegfried 8, 86–87, 257
Krebs, Paula M. 233, 257
Kyle of Lochalsh 149, 287

Lafayette, The Great (Sigmund Neuberger) 152, 287
Lancashire 118, 141
Landau, Paul S. 229, 257, 267
Lane, Edward 238
Lane, Tony 122, 124, 257
Lang, Fritz 17
Langfier, Adolphe 76, 276
Langton, John 6, 257, 266, 268
Larkin, Jim 176
Lauste, Emile 201, 236, 294–296
Ledger, Sally 257, 267
Leeds 13, 115–117, 285
Leersen, Jeop 172–173, 257
Leicester 117, 214, 284
leisure 24, 27, 31, 62, 66, 118–119
Leno, Dan 27
Levell, Nicky 92, 257
Lewis (Hebrides) 149, 287
Limehouse 17
Limerick 171, 182
lithography 54, 239–240
Liverpool 1, 10, 38–39, 55, 100, 111, 115–116, 118–119, 121–134, 139, 145, 147, 154–158, 161–163, 170, 230, 245–246, 256–257, 259, 264, 269, 271–273, 277, 282–283, 285
Lomas, Harold M. 94, 222, 237, 276, 278–280, 293–295
London 43–111, 275–281
London Cinematograph Company 174, 289
London Zoological Gardens 1, 85, 89–92, 95, 187, 249, 276, 279–281
Loubet, Émile 81, 280
Low, Rachel 20, 49, 189–190, 257, 262

Lowe, Donald 97, 257
Lowry, Donal 229, 257
Luckhurst, Roger 257, 267
Lucknow 210
Lumière, Auguste and Louis 8, 13, 43–44, 54–55, 76, 78, 91, 110, 168, 273, 275–282, 289–291

Macao 218
MacKenzie, Hugh 152, 258
MacKenzie, John M., vii 26, 28, 31–35, 47, 155, 157, 159, 194, 196–197, 209, 239, 267, 269
Madras 206, 255, 292
Makepeace, Chris 135, 258
Malins, Geoffrey 245
Manchester 1, 10, 26, 29, 32, 35–36, 38–39, 81, 86, 103, 111, 115–116, 118, 124, 129, 131–132, 134–145, 147, 154–156, 161–163, 165, 169, 191, 197, 202–203, 229–230, 234, 249–259, 261, 263–269, 283–284
Mansion House 56–58, 67–68, 75–76, 275–278
Mantle, Gregory J. 206, 292
Marble Arch 53, 55, 60, 68, 110, 215, 278
Marcus, Laura 258
markets 3, 6, 21, 25, 27, 33, 56, 70–72, 83, 117–118, 135–138, 140, 145, 177, 189, 205, 223, 235, 237, 253, 277–278, 280, 291, 293–295
Marriott, John 203, 258
Martins, Luciana 192–193, 253, 264
Maskelyne, J. Neville 77
Maver, Irena 156, 253, 268–269
Mayhew, Henry 5, 48, 60, 141, 258
Mazumdar, Ranjani 258
Mazzetti, Lorenza 245
McCaffrey, John F. 147, 159, 258
McCole, Niamh 171, 267
McDowell, J.B. 57, 61, 72, 117–118, 245, 276, 281, 285
McDowell, John K. 161, 258
McKernan, Luke 59, 176, 258
Melbourne 201, 296
Méliès, Georges 44
Mellin's Food 68, 142
Mellor, Robert 152, 267
Melsom, Sydney 230, 232, 294
Mersey 122–126, 128, 256, 282
Michie, Helena 105, 258, 270
Midlands 126
Midwinter, Eric 131, 259
military 19, 25–28, 31, 34, 52–53, 58, 63, 73, 75, 80, 118, 120–121, 130, 132, 134, 139, 141–142, 148, 154, 157, 166, 169, 179, 198–200, 203, 206, 229–234, 238, 241–244, 249, 258, 266
Mitchell, Sagar 22–24, 118–119, 126–133, 135, 138–144, 148, 150–151, 157–158, 160, 165, 170–173, 177–182, 230–231, 282–284, 286–291
Mitchell, Timothy 87–88, 182, 238, 259
Mitchell, William J.T. 12, 192, 259
modernization 48, 89, 98, 100, 104, 106, 110, 177–178, 224
Mogador 237, 295
Moholo-Nagy, László 17
monarchy 11, 47, 55, 59, 64, 75, 175, 264
Monks, Robert 170, 268
Morgan, Nick F. 156, 268
Morocco 194, 237, 296
Morris, Robert J. 6, 98, 123, 125, 159, 254, 257, 259, 264–266, 268–269
Moto-Photo Supply Co. 77
Motorgraph 77
Mullaney, Steven 73, 259
Mundy, Jennifer vii 55
municipalism 52, 54, 57, 63, 65, 103–104, 127, 143, 145, 151, 155, 157, 159, 179, 212, 265, 278, 288
Munro, Ian 73, 259
Murray, John 153, 194, 229, 250, 255–256

music halls 2, 21, 26–28, 32–35, 44–45, 50, 78, 80, 90, 94, 106, 171, 191, 193, 213–214, 230–231, 269, 273
Musser, Charles 9–11, 259

Nanjing 218, 223–224, 252, 293
Nash, John 89–90
Natal 75, 229, 236
nationalism 12, 25–28, 34, 131, 134, 143, 162–163, 166, 168, 174–175, 179, 181, 207, 230, 249, 251
Nead, Lynda vii 23, 25–26, 47–48, 67, 258–259, 271
Neal, Frank 131, 259
Nelson, Horatio 65, 126, 132
Nestlé 68–69
New Zealand 139, 201
Newcastle 126, 282
Newhall, Beaumont 239, 259
newspapers 2, 33, 43, 45, 54, 80, 90, 168, 209–210, 218, 231, 272
Nicholls, David 138, 256, 265
Nichols, Bill 7–8, 259
Nicholson, John 4, 39, 210–211, 255, 257
Nickelodeon 27, 271
Nightingale, Florence 63
Noble, C. Rider 230, 232, 237, 240, 294, 296
North American Animated Photo Company 170, 178
Northern Photographic Works 169, 175, 275–276, 281, 290, 296
Norton, Charles G. 72, 101, 105–107, 250, 277–278, 280
Notari, Elvira 15, 251
Nottingham 117, 285

O'Keefe, John 197
Oakley, Charles 20, 161, 259
Obama 149
Oban 149, 287
Okehampton 120
Oksilof, Assenka 243, 259
Olsson, Jan 17, 253, 265
Omdurman 230, 294
omnibuses 64–65, 67, 69, 82
Opium War 219
Orientalism 34–35, 157, 159, 166, 198–199, 220, 240, 249–250, 254, 258, 261, 269
Ormiston-Smith, F. 237, 295
Ornamentalism 30, 211, 251
Orwell, George 3, 259
Osterhammel, Jürgen 218–219, 224, 268
Ové, Horace 246

Pagani, Catherine 219, 268
pageantry 34, 62, 78, 94, 117, 130, 151, 215, 268, 276, 279, 282, 285–286
painting 2, 16, 23, 33, 51, 55, 62, 70, 75, 88, 108, 121, 149, 169, 190–192, 197, 203, 207, 259
Pakenham, Thomas 229, 259
Palestine 237, 239
Palmberg, Mai 227, 259
Palmer, Sarah 122–123, 268
Panofsky, Erwin 37, 69, 259–260
Park, Mungo 193
parks 1, 43–45, 47, 49, 53, 55–60, 65, 74, 89–95, 100, 104, 106, 125, 143–144, 148, 153, 157–158, 160, 175, 179, 181, 246, 275–281, 284, 287–288, 290
Parnell, Charles S. 168
Pathé 56, 59, 91, 150, 171, 176, 208, 214, 216, 223, 275, 278, 289, 291–292
patriotism 25, 27–28, 32–33, 71, 78, 98, 132–133, 202, 210, 218, 230–231, 261, 265, 268–269, 272
Paul, Robert W. 19, 24, 26, 43–45, 51, 53–54, 57–58, 61–64, 70, 72–73, 75–81, 86, 101–102, 107, 117–121, 126, 153–154, 169, 171, 189, 206, 213–214, 230, 232, 234, 271,

275–282, 284–287, 289, 291–292, 294–295
pauperism 141
pavements 104–106, 132, 138, 153, 161, 210
pedestrians 13, 51, 61, 64–65, 105, 107, 117, 159–160, 169, 179, 204, 235
Peking 217, 220–223, 242–243, 292–293
Penz, François vii 16, 260, 270
Perry, George 20, 260
Perth 148, 288
Peshawar 206, 292
Pevsner, Nikolaus 68, 251
Photogram 76–77, 273
photography 2, 5, 7, 11, 13, 16, 18, 20, 23–24, 31, 33, 36–37, 43, 47–49, 54, 62, 64, 73, 77–78, 86, 88, 101–102, 108–109, 115, 135, 148–149, 156–158, 169, 171–172, 175, 188–191, 194–197, 203, 206, 211, 214, 216, 219, 222, 228–229, 231–232, 238–240, 249, 252, 258–259, 261, 265, 267, 269, 275–276, 281, 286, 290, 292, 296
Picard, Liza 187, 260
Pictorialograph 56, 77, 278
Plomer, William 124, 260
Plymouth 120, 285
Poe, Edgar Allan 10, 70, 260
Poona 205–206, 292
Popple, Simon 22, 135, 230–231, 262, 266, 268, 270
Port Elizabeth 232–236, 295
Port Said 240–241, 294
Porter, Andrew 28, 167, 219, 265, 267–268
Porter, Bernard 30, 260
Porter, Laraine 23, 103, 266
Porter, Roy 48, 197, 261, 267–268
Porter, Vincent 245, 254
Portsmouth 118, 120–121, 285
postcolonialism 166, 174, 193, 229, 251, 257, 260, 265, 267, 269
poverty 5, 38, 48, 58, 72, 74, 140–141, 157, 206
Prescott, Andrew 144, 268
Prestwich Manufacturing Company 76, 78
Price, Richard N. 24–25, 230, 260, 268
Prince, Louis Le 13, 115–117, 128, 132, 143, 151, 182, 243, 282, 285, 291
Pringle, Ralph 126–128, 130, 178–179
Promio, Alexandre 54–55, 59, 61–63, 90–91, 93, 121, 125–126, 128–129, 134, 168–170, 178–179, 193, 271, 275–282, 289–291
propaganda 26–29, 34, 47, 53, 74, 194, 196–197, 202, 213, 215, 227–228, 233, 242–243, 245, 258, 260
prostitution 60, 263
Przyblyski, Jeannene M. 99, 261, 269

Queen Victoria's Diamond Jubilee 2, 47, 49, 68, 70, 74–80, 83, 109, 130, 132, 208–209, 212, 214–215, 250, 280
Quilley, Geoff 191, 250
Quiribet, Gaston 62, 65, 68, 102, 276, 279, 281

R.W. Paul's Theatrograph 117, 120, 126, 232, 240
Raack, Richard C. 6–7, 271
railway 1, 97, 99–102, 105, 107–108, 120, 122–123, 125–126, 134, 137, 139, 148–149, 161, 177, 179, 181, 212, 237, 257, 261, 276–278, 280, 282, 285–288, 291
railway stations 1, 56, 64, 66, 77, 97, 99–101, 107, 120, 135, 142, 148, 154, 169, 177, 206, 210, 232, 237, 276–278, 286, 289, 291–292
rallies 25, 58–59, 65
Ranger, Terrence 11, 209, 255, 264
Reeves, Nicholas 260
Reisz, Karel 245

Richards, Jeffrey 166–167, 268
Richards, Thomas 36, 75–76, 86, 95, 190, 255, 260
rickshaws 217, 224, 236, 294
Rigier, Bernhard 252, 266
Riley Bros. 117–118, 284–285
rivers 52, 62, 64, 123–126, 155, 157, 160, 180, 207, 219, 223, 233, 241, 255, 282, 293
Roberts, David 239
Roberts, Frederick Sleigh vii 60, 79, 120, 131–132, 142–143, 147, 153, 229–230, 232–233, 278, 282–283, 288, 294
Roberts, Kenneth W. 256, 267
Roberts, Mary 199, 254
Robinson, Ronald 238, 261
Rockett, Kevin 168–169, 171, 255, 261, 264, 267
Rodger, Richard 6, 98, 150, 159, 259, 261, 265, 269
Rosenthal, Joe 126, 200–201, 223–224, 230, 232, 240, 282, 292–294, 296
Ross, Robert 199, 261, 267
Rousseau, George S. 197, 261, 267–268
Rowe, J.W. 56, 77, 278
Rowe, Josiah 194
Royal Exchange 51, 140
Russell, Patrick 22, 135, 262, 266, 268, 270
Ruttmann, Walter 245
Ryan, Deborah S. 94, 268
Ryan, James R. 36, 195, 239, 261, 265, 269

Sadoul, Georges 20, 118, 261, 271
Said, Edward 34–35, 47, 87, 166, 198, 240–241, 261, 269, 294
sailing 52, 120–121, 124–125, 207, 217
Saint, Andrew 64, 265
Salford 135, 140, 143, 252, 258, 283
Salisbury 120, 285
Sandakan 222
Sanders, James 261
Schivelbusch 99, 261, 269
Schmiechen, James 269
Schneer, Jonathan 28, 36, 52–53, 67, 90, 261
Schwartz, Joanna 195, 239, 261, 265
Schwartz, Vanessa 4–5, 18, 45, 99, 261, 269
Schwarze, Tracey T. 174, 272
Scotland v 38, 56, 118, 126, 147–157, 159, 161, 163, 258, 286–289
Scott, Fred 5, 140–141
seaside v, 1, 115, 117–119, 121, 123, 125, 127, 129, 131, 133, 135, 137, 139, 141, 143, 145, 263
Second Boer War vi 27–28, 52, 60, 63, 81, 96, 120, 131–132, 142–143, 153, 161, 165–166, 201, 206, 221, 224–225, 227, 229–237, 239, 241–245, 247, 250, 256–257, 259–260, 265, 268, 282, 291
sectarianism 123, 131, 156, 162, 269
Seeley, John Robert 1, 261
Sennett, Richard 89, 116, 204, 261
Sepoy Mutiny 1, 208–211, 214–216, 218, 231, 292
settlements 200–201, 212, 222, 224, 228
sewage system 122, 129, 136, 156, 162
Shakespeare, William 69
Shandong 220
Shanghai 201, 218, 222–224, 252, 293
Shanxi 220
Shepherd's Bush (London) 103–104, 276, 281
Shiel, Mark 14–15, 39, 262, 269
Shields, Ella 27
shipbuilding 155–156
Shipman, David 20, 262
ships 126–128, 134, 139–140, 158, 163, 179, 209, 284
shops 53, 70, 83, 104, 106, 120, 128–129, 136–138, 155, 169, 180, 194, 223–224, 235, 286, 293

Sickert, Walter A. 27
Sydney 20, 63, 125, 254, 264, 296
Simms, Anngret 173, 251
Simpson, William 203
Singapore 201, 217–218, 222–223, 293
Sjöström, Victor 17
Skladanowsky, Max and Emil 13
slums 5, 18, 47–49, 52, 57, 105–106, 123–125, 136, 138, 141, 145, 154, 156, 162, 181, 234, 245, 250, 280
Smith, George Albert 20, 24, 44, 61, 65, 76, 81, 93–94, 119–120, 227, 230, 275–281, 284–285, 295
Smith, Joan 123, 131, 156, 162, 269
Smith, John S. 152, 267
Smithfield 72, 278
socialism 25, 155, 157, 159, 246, 265
Soja, Edward W. 14, 29, 262
soldiers 12, 28, 52, 72, 75, 98, 120, 141–142, 151, 153, 157, 167, 179, 209, 211, 221, 232, 241, 285
Southampton 26, 118, 120–121, 123, 128, 131, 158, 209, 230, 285–286
Spitalfields 72, 280
sports 19, 118, 126–127, 143, 148, 151, 155, 213, 282, 286–287, 289
Sprinker, Michael 35, 262
squares 1, 30, 49, 53–54, 56, 58–59, 61–68, 76, 82, 102, 136–137, 142, 150, 156, 159, 161, 173, 187, 210, 218, 228, 237, 240, 258, 276, 278, 281, 294–295
steamers 62–63, 124–129, 158–159, 282
Stevenson, David 152, 262, 267
Steward, F.B. 292
Stoker, Bram 50, 262
Stollery, Martin 31, 262
Storch, Robert 32, 35, 262, 269
Strand (London) 53–56, 58, 64, 66–67, 70, 73, 76–77, 244, 278, 295
Strand, Paul 244
Stranraer 148, 287
strikes 6, 150, 176, 289
Studlar, Gaylyn 35, 250
suburbs 6, 87, 97–98, 100, 103–106, 108, 122, 129, 138, 163, 181, 212, 269
Suffragettes 59, 65, 280–281
Summerfield, Penny 32, 269
Surridge, Keith 229, 256
Sutcliffe, Anthony 15, 262
Sutherland 149
Swinton, William 262
Sydenham 92, 266
Sydney 201, 294

Tangier 237, 294–295
Tanner, Alain 245
Taylor, Richard 44, 251
Telkamp, George J. 199, 261, 267
Thames 52, 56, 61–63, 103, 276–277, 279, 281
Thomas, Nicholas 26, 192, 198
Thompson, Andrew S. 30, 272
Thompson, Francis M. 98, 269
Thompson, John 5, 49, 156, 219–220, 262
Thompson, John H. 200, 270
Thompson, Kristin 20, 250
thoroughfare 5, 8, 13, 62, 67, 70, 127, 129, 145, 169, 201, 224, 236, 244
Tiedemann, Rolf 244, 250, 263
Tientsin 221, 223, 292–293
Tomaselli, Keyan 5, 262
Topical Film Company 177, 291
Toulmin, Vanessa vii 22–23, 118–119, 126–127, 129, 132, 135, 142, 144, 170, 262, 266, 268, 270
tourism 5, 53, 63, 110, 118–119, 191, 238, 265
Townshend, Charles 166, 174, 262
trade 115, 117, 121–122, 125, 127, 129–130, 134, 137, 139, 143, 148, 163, 176–177, 182, 218–219, 238, 244, 266, 287
Trafalgar Day (Liverpool) 130, 132–133, 283

Trafalgar Square 53–54, 56, 63–66, 76, 82, 258, 272, 278, 281
traffic 9–10, 51, 55–57, 63, 66–68, 70, 100–101, 103, 105–109, 115, 117, 123, 138, 145, 160, 162, 180, 210, 217, 235, 271, 275–278, 280–281
trains 1, 13, 23, 32, 87, 99–103, 125–126, 148, 151, 159, 168–169, 177, 192, 206, 210, 237, 276, 280, 282, 288–289, 291–292, 294
trams 23, 61, 65, 68, 87, 99, 102–109, 117, 119–121, 128–129, 132, 134, 138, 142, 144–145, 148, 159–162, 173, 176, 178–180, 212, 223, 234–236, 266, 276–277, 279, 281, 286, 289–291, 293–296
travelogues 56, 149, 170, 197, 203, 208–209, 242
Trevithick, Alan 212–213, 272
troops 25–26, 33, 47, 74, 76, 78–81, 83, 120, 131–132, 141–142, 151, 157, 176, 211, 221, 223–224, 229–230, 232–234, 241, 292–293, 295
Turner, Florence 149, 289
Twain, Mark 207
Tyler, Walter 57, 126, 277, 282

uniforms 2, 35, 80, 132–133, 143, 236
Urban, Charles 24, 44, 56, 60, 62, 68, 72, 81, 91, 94, 96, 100, 117, 120–121, 149–150, 154, 169–171, 174, 206, 212, 214, 217, 222, 237, 242, 275–281, 285–289, 291–293, 295–296
Urban Trading Company 242, 275–281, 285–289, 292–293, 295–296
urbanization 6, 32, 46, 103, 106, 109, 147

Vadillo, Ana 105, 270
Varanasi 187, 201, 206–208, 216
Varley, Frederick 57, 115
Velograph Syndicate 76–78, 105
Vertov, Dziga 13, 244
Victoria and Albert Museum 95–96, 208, 217, 277–278
Victoria, queen of England 1, 11, 26, 33, 47, 49, 51, 54, 56, 62, 67–70, 74–76, 78–81, 95–96, 99–100, 109, 117, 130, 142, 144, 153, 157, 165–166, 175, 200, 208, 214–215, 239, 258–259, 269, 277–278, 280, 290–291
Vidler, Anthony 16–17, 262, 272
Vitagraph 55, 59, 76, 80, 96, 120, 125–126, 174, 180, 277–279, 283, 285, 290
Volwahsen, Andreas 263

W. Watson & Sons 65, 77, 132, 281
Wales 80, 127, 143, 231–232, 282, 295
Walkowitz, Judith R. 60, 263
Walton, John K. 118, 263
Walturdaw 68, 91, 96, 119, 204, 227, 275, 279–281, 284, 292, 295
Wargraph 231
Warwick Trading Company 47, 55–57, 59–60, 62–63, 65, 76, 81, 91, 93, 96, 102–103, 117, 119–120, 126–127, 131–132, 135, 139, 148, 151, 169, 171, 174–175, 178, 180, 189, 200, 205–206, 213–214, 223, 227–228, 230, 232, 234, 237, 240, 276–285, 287–289, 291–296
Waterford 72, 182
Waterloo Place 56, 58, 63, 278
waterworks 64
Webb, Aston 95
Weber, Andrew 23, 267
Wegener, Paul 16
Weihsmann, Helmut 9, 16–17, 266, 270
Weller, Edward 263
Westminster Abbey 54, 61, 81–83, 281
Weston 118, 284
Westport 32, 253

White, Jerry 59, 83, 263
Whitechapel 57, 71, 129, 280
Whitehall 53, 56, 64–65, 278
Wiene, Robert 16
Williams, Raymond 73–74, 263
Williamson, James 20, 33, 82, 119, 151, 221–222, 276, 285, 288
Wilson, Emma 23, 267
Withers, Charles W.J. 256
Wolf, Philipp 117, 286
Wolverhampton 117, 286
Wood, Daisy 27
Woodside (Mersey) 125–126, 283
Woodstock (South Africa) 234, 296
Worcester 117, 286
workers 26, 100, 121–122, 124, 126–128, 135–136, 148, 176, 229–230, 284, 289
Wrech, Alfred 77
Wright, Basil 101

Yearsley, Ian 23, 270
Yorkhill (Glasgow) 157, 287
Yorkshire 118

Zimbabwe 227
Ziter, Edward 191, 236–237, 263

NEW STUDIES IN EUROPEAN CINEMA

With its focus on new critical, theoretical, and cultural developments in contemporary film studies, this series develops rigorous analytical debate within an innovative, multidisciplinary, and transnational approach to European cinema.

Through a mixture of edited collections and single-authored volumes, the series aims both to re-evaluate established critical thought and to identify and explore new trends and theories that will inform cinema studies over the coming decades. It provides an international forum for lively and controversial debate embracing all aspects of European cinema from a broad range of theoretical perspectives. The *New Studies in European Cinema* series thus makes a key contribution to a subject whose importance to contemporary culture and identity is fundamental.

Vol. 1 William Hope (ed.)
Italian Cinema. New Directions.
2005. ISBN 3-03910-282-6 / US-ISBN 0-8204-7209-3

Vol. 2 Wendy Everett & Axel Goodbody (eds)
Revisiting Space. Space and Place in European Cinema.
2005. ISBN 3-03910-264-8 / US-ISBN 0-8204-7193-3

Vol. 3 David Montero
Thinking Images. The Essay Film as a Dialogic Form in European Cinema.
2012. ISBN 978-3-0343-0730-7

Vol. 4 Ewa Mazierska
Polish Postcommunist Cinema. From Pavement Level.
2007. ISBN 3-03910-529-9 / US-ISBN 0-8204-7533-5

Vol. 5 Forthcoming

Vol. 6 Wendy Everett (ed.)
Questions of Colour in Cinema. From Paintbrush to Pixel.
2007. ISBN 978-3-03911-353-8

Vol. 7 Rosanna Maule (ed.) with Julie Beaulieu
In the Dark Room: Marguerite Duras and Cinema.
2009. ISBN 978-3-03911-354-5

Vol. 8 Leila Wimmer
Cross-Channel Perspectives. The French Reception of British Cinema.
2009. ISBN 978-3-03911-360-6

Vol. 9 Lucy Bolton and Christina Siggers Manson (eds)
Italy on Screen. National Identity and Italian Imaginary.
2010. ISBN 978-3-03911-416-0

Vol. 10 Matthias Uecker
Performing the Modern German. Performance and Identity in Contemporary German Cinema.
2013. ISBN 978-3-0343-0972-1

Vol. 11 Rosemary Stott
Crossing the Wall. The Western Feature Film Import in East Germany.
2012. ISBN 978-3-03911-944-8

Vol. 12 Pietari Kääpä
The National and Beyond. The Globalisation of Finnish Cinema in the Films of Aki and Mika Kaurismäki.
2010. ISBN 978-3-03911-966-0

Vol. 13 Ben McCann
Ripping Open the Set. French Film Design, 1930–1939.
2013. ISBN 978-3-03910-311-9

Vol. 14 Fiona Handyside
Cinema at the Shore. The Beach in French Film.
2014. ISBN 978-3-0343-0834-2

Vol. 15 Maurizio Cinquegrani
Of Empire and the City. Remapping Early British Cinema.
2014. ISBN 978-3-0343-0823-6

www.ingramcontent.com/pod-product-compliance
Ingram Content Group UK Ltd.
Pitfield, Milton Keynes, MK11 3LW, UK
UKHW020201290926
15471UKWH00005B/998

9 783034 30823